AF878414

Raspberry Pi Pico W

Program, build, and master 60+ projects with the **Wireless RP2040**

Dogan Ibrahim

elektor
design > share > sell

This is an Elektor Publication. Elektor is the media brand of
Elektor International Media B.V.
PO Box 11, NL-6114-ZG Susteren, The Netherlands
Phone: +31 46 4389444

Declaration

The Author and Publisher have used their best efforts in ensuring the correctness of the information contained in this book. They do not assume, and hereby disclaim, any liability to any party for any loss or damage caused by errors or omissions in this book, whether such errors or omissions result from negligence, accident, or any other cause.

All the programs given in the book are Copyright of the Author and Elektor International Media. These programs may only be used for educational purposes. Written permission from the Author or Elektor must be obtained before any of these programs can be used for commercial purposes.

British Library Cataloguing in Publication Data
A catalogue record for this book is available from the British Library

ISBN 978-3-89576-529-2 Print
ISBN 978-3-89576-530-8 eBook

Editor: Alina Neacsu
Prepress Production: D-Vision, Julian van den Berg

Contents

Preface

A microcontroller is basically a single-chip computer including a CPU, memory, input-output circuitry, timers, interrupt circuitry, clock circuitry, and several other circuits and modules, all housed in a single silicon chip. Early microcontrollers were limited in their capacities and speed and they consumed considerable power. Most of the early microcontrollers were 8-bit processors with clock speeds in the region of several megahertz, and having only hundreds of bytes of program and data memories. These microcontrollers were traditionally programmed using assembly languages of the target processors. Today, 8-bit microcontrollers are still in common use, especially in small projects where large amounts of memory or high speed are not the main requirements. With the advancement of chip technology, we now have 32-bit and 64-bit microcontrollers with speeds in the region of several gigahertz, and several gigabytes of memory. Microcontrollers are nowadays programmed using high-level languages such as C, C#, BASIC, PASCAL, JAVA, etc.

The Raspberry Pi Pico is a high-performance microcontroller designed especially for physical computing. Readers should realize that microcontrollers are very different from single-board computers like the Raspberry Pi 4 (and other family members of the Raspberry Pi). There is no operating system on the Raspberry Pi Pico. Microcontrollers like the Raspberry Pi Pico can be programmed to run a single task and they can be used in fast real-time control and monitoring applications.

The Raspberry Pi Pico is based on the fast and very efficient dual-core ARM Cortex-M0+ RP2040 microcontroller chip, running at up to 133 MHz. The chip incorporates 264 KB of SRAM and 2 MB of Flash memory. What makes the Raspberry Pi Pico very attractive is its large number of GPIO pins, and commonly used peripheral interface modules, such as SPI, I^2C, UART, PWM, and fast and accurate timing modules.

Released in 2022, the Raspberry Pi Pico **W** is the latest model of the Pico family of microcontroller boards. The "Pico W" is identical to the standard "Pico" with one major difference: compared to other Pico family members, the "Pico W" has an on-board Wi-Fi module, thus enabling the board to be used in many communications, control, and especially in IoT based projects. The Pico W board also supports Bluetooth hardware, but the Bluetooth firmware was not ready at the time of writing this book.

Both the standard Raspberry Pi Pico and the Raspberry Pi Pico W can easily be programmed using some of the popular high-level languages, such as MicroPython, or C/C++. There are many application notes, tutorials, and data sheets available on the Internet on using the Pico or the Pico W.

This book is an introduction to using the Raspberry Pi Pico W microcontroller development board with the MicroPython programming language. The Thonny development environment (IDE) is used in all the projects in the book and the readers are recommended to use this IDE. There are many working and tested projects in the book, covering almost all aspects of the Raspberry Pi Pico W. Excepting the Wi-Fi-based subjects, all projects in the book can also be used with the standard Raspberry Pi Pico without any modifications.

The following sub-headings are given for each project where applicable to make the projects easy to follow for the readers:

- Title
- Brief description
- Aim
- Block diagram
- Circuit diagram
- Program listing with full description

I hope your next microcontroller-based projects make use of the Raspberry Pi Pico W, and that this book becomes useful in the development of your projects.

Dr Dogan Ibrahim
London, 2022

Chapter 1 • Raspberry Pi Pico W Hardware

1.1 Overview

The Raspberry Pi Pico W is a single-board microcontroller module developed by the Raspberry Pi Foundation. This module is based on the RP2040 microcontroller chip. In this chapter, you will be looking at the hardware details of the Raspberry Pi Pico W microcontroller module in some detail. From now on, this microcontroller module Pico W will be called "Pico" for short.

1.2 The Pico hardware module

Pico is a very low-cost $6 microcontroller module based on the RP2040 microcontroller chip with a dual-core Cortex-M0+ processor. Figure 1.1 shows the front view of the Pico hardware module which is basically a small board. In the middle of the board is the tiny 7 × 7 mm RP2040 microcontroller chip housed in a QFN-56 package. At the two edges of the board, there are 40 gold-coloured metal GPIO (General-Input-Output) pins with holes. You should solder pins to these holes so that external connections can be made easily to the board. The holes are marked starting with number 1 at the top left corner of the board and the numbers increase downwards up to number 40 which is at the top right-hand corner of the board. The board is breadboard compatible (i.e., 0.1-inch pin spacing), and after soldering the pins, the board can be plugged on a breadboard for easy connection to the GPIO pins using jumper wires. Next to these holes, you will see bumpy circular cut-outs which can be plugged in on top of other modules without having any physical pins fitted.

Figure 1.1: Front view of the Pico hardware module.

At one edge of the board, there is the micro-USB B port for providing power to the board and for programming the board. Next to the USB port, there is an on-board user LED that can be used during program development. Next to this LED, there is a button named BOOTSEL that is used during the programming of the microcontroller as you will see in the next chapters. Next to the processor chip, there are 3 holes where external connections can be made to. These are used to debug your programs using Serial Wire Debug (SWD). At the other edge of the board, is the single-band 2.4 GHz Wi-Fi module (802.11n). next to the Wi-Fi module is the on-board antenna.

Figure 1.2 shows the back view of the Pico hardware module. Here, all the GPIO pins are identified with letters and numbers. You will notice the following types of letters and numbers:

GND	- power supply ground (digital ground)
AGND	- power supply ground (analog ground)
3V3	- +3.3 V power supply (output)
GP0 – GP22	- digital GPIO
GP26_A0 – GP28_A2	- analog inputs
ADC_VREF	- ADC reference voltage
TP1 – TP6	- test points
SWDIO, GND, SWCLK	- debug interface
RUN	- default RUN pin. Connect LOW to reset the RP2040
3V3_EN	- this pin by default enables the +3.3 V power supply. +3.3 V can be disabled by connecting this pin LOW
VSYS	- system input voltage (1.8 V to 5.5 V) used by the on-board SMPS to generate +3.3 V supply for the board
VBUS	- micro-USB input voltage (+5 V)

Figure 1.2: Back view of the Pico hardware module.

Some of the GPIO pins are used for internal board functions. These are:

GP29 (input)	- used in ADC mode (ADC3) to measure VSYS/3
GP24 (input)	- VBUS sense - HIGH if VBUS is present, else LOW
GP23 (output)	- Controls the on-board SMPS Power Save pin

The specifications of the Pico hardware module are as follows:

- 32-bit RP2040 Cortex-M0+ dual-core processor operating at 133 MHz
- 2 MByte Q-SPI Flash memory
- 264 KByte SRAM memory
- 26 GPIO (+3.3 V compatible)
- 3× 12-bit ADC pins
- Accelerated floating point libraries on-chip

- On-board single-band Infineon CYW43439 wireless chip, 2.4 GHz wireless interface (802.11b/g/n) and Bluetooth 5.2 (not supported at the time of writing)
- Serial Wire Debug (SWD) port
- Micro-USB port (USB 1.1) for power (+5 V) and data (programming)
- 2× UART, 2 × I²C, 2 × SPI bus interface
- 16 PWM channels
- 1× Timer (with 4 alarms), 1× Real Time Counter
- On-board temperature sensor
- On-board LED at GPIO0, controlled by the 43439 module
- Castellated module allowing soldering direct to carrier boards
- 8× Programmable IO (PIO) state machines for custom peripheral support
- MicroPython, C, C++ programming
- Drag & drop programming using mass storage over USB

Notice that on the Raspberry Pi Pico board, the on-board LED is connected to GP25 and is available to the user. On the Raspberry Pi Pico W board, the on-board LED is controlled by the 43439 Wi-Fi module.

Pico GPIO hardware is +3.3 V compatible, and it is, therefore, important to be careful not to exceed this voltage when interfacing external input devices to the GPIO pins. +5 V to +3.3 V logic converter circuits or resistive potential divider circuits must be used if it is required to interface devices with +5 V outputs to the Pico GPIO pins.

Figure 1.3 shows a resistive potential divider circuit that can be used to lower +5 V to +3.3 V.

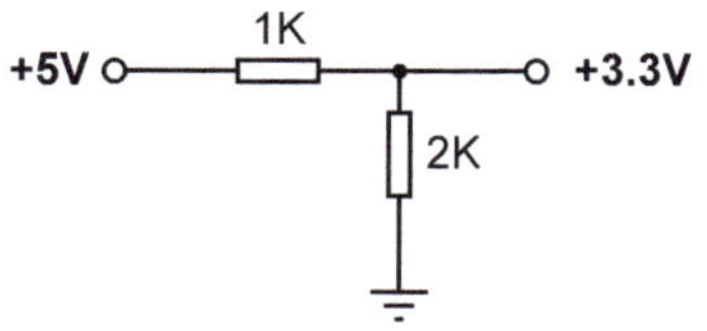

Figure 1.3: Resistive potential divider circuit.

1.3 Comparison with the Arduino UNO

Arduino UNO is one of the most popular microcontroller development boards used by students, practising engineers, and hobbyists. Table 1.1 shows a comparison of the Raspberry Pi Pico W with the Arduino UNO. It is clear from this table that the Pico W is much faster than the Arduino UNO, it has larger flash and data memories, offers Wi-Fi, provides more digital input-output pins, and has an on-board temperature sensor. Arduino UNO operates with +5 V and its GPIO pins are +5 V compatible. Perhaps some advantages of the Arduino UNO are that it has a built-in EEPROM memory, and its ADC is 6 channels instead of 3.

Feature	Raspberry Pi Pico W	Arduino UNO
Microcontroller	RP2040	Atmega328P
Core and bits	Dual core, 32-bits, Cortex-M0+	Single-core 8-bits
RAM	264Kbyte	2KByte
Flash	2MByte	32KByte
CPU speed	48MHZ to 133MHz	16MHz
EEPROM	None	1KByte
Wi-Fi support	CYW43439 wireless chip	None
Power input	+5V through USB port	+5V through USB port
Alternative power	2 – 5V via VSYS pin	7 – 12V
MCU operating voltage	+3.3V	+5V
GPIO count	26	20
ADC count	3	6
Hardware UART	2	1
Hardware I^2C	2	1
Hardware SPI	2	1
Hardware PWM	16	6
Programming languages	MicroPython, C, C++	C (Arduino IDE)
On-board LED	1	1
Cost	$6	$20

Table 1.1: Comparison of Raspberry Pi Pico W and Arduino UNO.

1.4 Operating conditions and powering the Pico

The recommended operating conditions of the Pico are:

- Operating temperature: -20 ^{o}C to +85 ^{o}C
- VBUS voltage: +5 V ±10%
- VSYS voltage: +1.8 V to +5.5 V

An on-board SMPS is used to generate the +3.3 V to power the RP2040 from a range of input voltages from 1.8 V to +5.5 V. For example, 3 alkaline AA batteries can be used to provide +4.5 V to power Pico.

Pico can be powered in several ways. The simplest method is to plug the micro-USB port to a +5 V power source, such as the USB port of a computer or a +5 V power adapter. This will provide power to the VSYS input (see Figure 1.4) through a Schottky diode. The voltage at the VSYS input is therefore VBUS voltage minus the voltage drop of the Schottky diode (about +0.7 V). VBUS and VSYS pins can be shorted if the board is powered from an external +5 V USB port. This will increase the voltage input slightly and hence reduce ripples on VSYS. VSYS voltage is fed to the SMPS through the RT6150 which generates fixed +3.3 V for the MCU and other parts of the board. VSYS is divided by 3 and is available at analog input port GPIO29 (ADC3) which can easily be monitored. GPIO24 checks the existence of VBUS voltage and is at logic HIGH if VBUS is present.

Another method to power the Pico is by applying external voltage (+1.8 V to +5.5 V) to the VSYS input directly (e.g., using batteries or an external power supply). You can also use the USB input and VSYS inputs together to supply power to Pico, for example, to operate with both batteries and the USB port. If this method is used, then a Schottky diode should be used at the VSYS input to prevent the supplies from interfering with each other. The higher voltages will power VSYS.

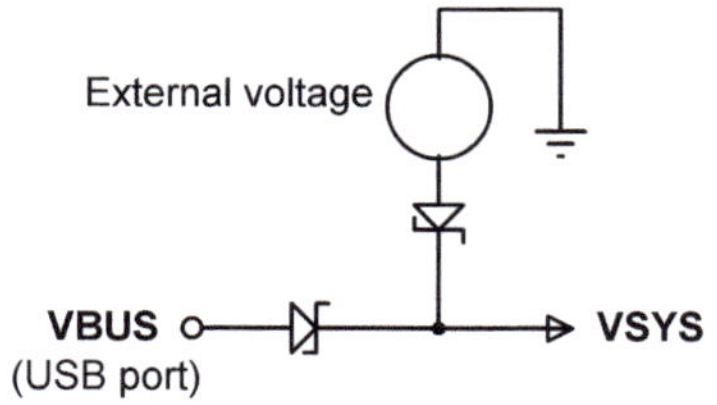

Figure 1.4: Powering the Pico.

1.5 Pinout of the RP2040 microcontroller and Pico module

Figure 1.5 shows the RP2040 microcontroller pinout, which is housed in a 56-pin package. The Pico module pinout is shown in Figure 1.6 in detail. As you can see from the figure, most pins have multiple functions. For example, GPIO0 (pin 1) is also the UART0 TX, I2C0 SDA, and the SPI0 RX pins.

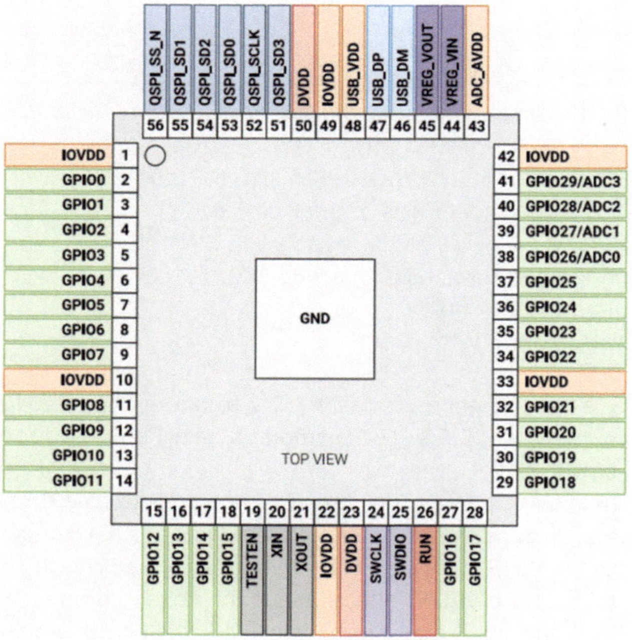

Figure 1.5: RP2040 microcontroller pinout.

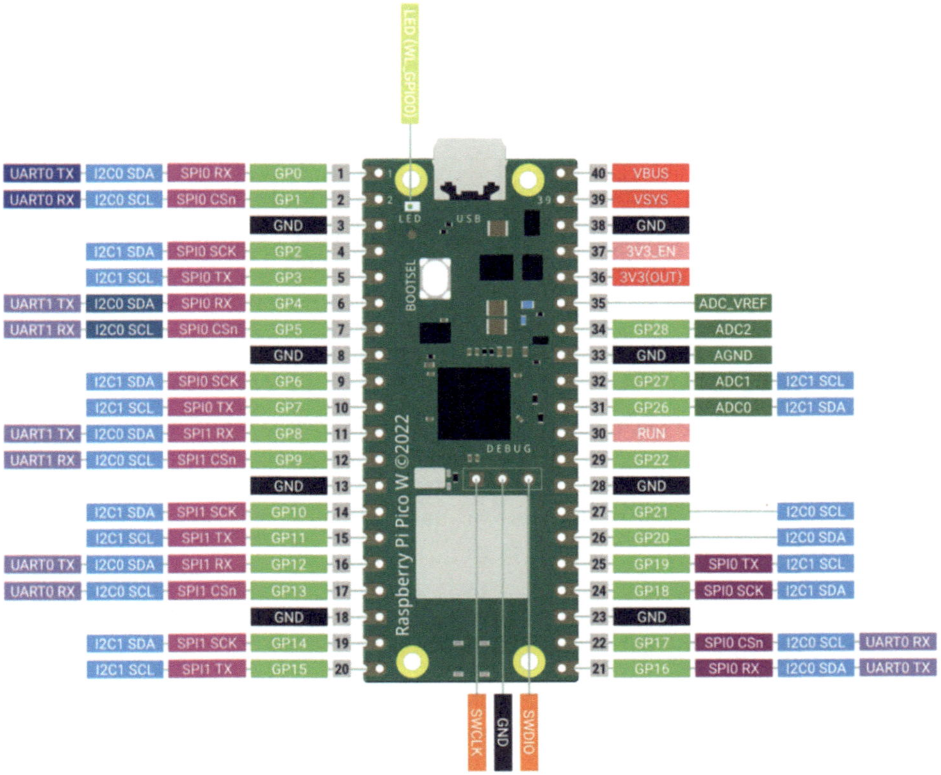

Figure 1.6: Pico pinout.

Figure 1.7 shows a simplified block diagram of the Pico hardware module. Notice that the GPIO pins are directly connected from the microcontroller chip to the GPIO connector. GPIO 26-28 can be used either as digital GPIO or as ADC inputs. ADC inputs GPIO26-29 have reverse diodes to 3Vs and therefore the input voltage must not exceed 3V3+300 mV. Another point to note is that if the RP2040 is not powered, applying voltages to GPIO26-29 pins may leak through the diode to the power supply (there is no problem with the other GPIO pins and voltage can be applied when the RP2040 is not powered).

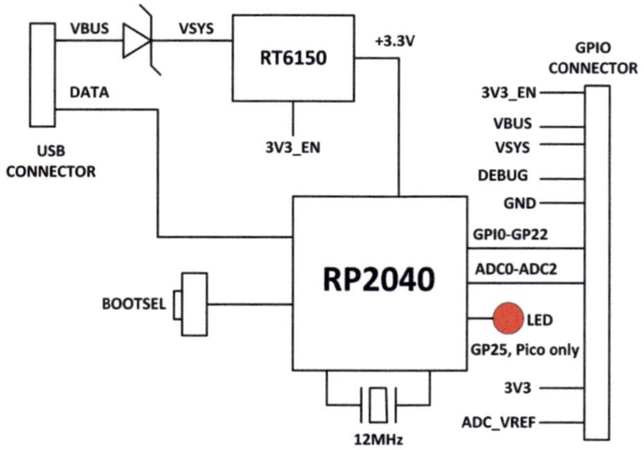

Figure 1.7: Simplified block diagram.

1.6 Other RP2040 microcontroller-based boards

There are many other RP2040 microcontroller-based development boards. At the time of writing this book some of these boards were:

- Adafruit Feather RP2040
- Adafruit ItsyBitsy RP2040
- Adafruit QT Py RP2040
- Pimoroni PicoSystem
- Pimoroni Tufty 2040
- Arduino Nano RP2040 Connect
- SparkFun Thing Plus RP2040
- Pimoroni Pico Explorer Base
- Pimoroni Pico Lipo
- SparkFun MicroMod RP2040 Processor
- SparkFun Pro Micro RP2040
- Pico RGB Keypad Base
- Pico Omnibus
- Pimoroni Pico VGA Demo Base
- Cytron Maker Pi RP2040 development board
- Technoblogy – Minimal RP2040 board
- Pimoroni Tiny 2040
- Seeed Studio Wio RP2040 Mini development board
- Seeed XIAO RP2040 development board
- And many more

Chapter 2 • Raspberry Pi Pico W Programming

2.1 Overview

At the time of writing this book, the Raspberry Pi Pico W accept programming with the following programming languages:

- C/C++
- MicroPython
- Assembly language

Although Pico by default is set up for use with the powerful and popular C/C++ language, many beginners find it easier to use MicroPython, which is a version of the Python programming language developed specifically for microcontrollers.

In this chapter, you will learn how to install and use the MicroPython programming language. You will be using the Thonny text editor which has been developed specifically for Python programs.

Many working and fully tested projects will be given in the next chapters using MicroPython with your Pico.

2.2 Installing MicroPython on Pico W

MicroPython must be installed on Raspberry Pi Pico W before the board can be used. Once installed MicroPython stays on your Pico unless it is overwritten with something else. Installing the MicroPython requires an Internet connection, and this is required only once. This can be done either by using a Raspberry Pi (e.g., Raspberry Pi 4), or by using a PC. In this section, you will see how to install using a PC (e.g., Windows 10).

The steps are as follows:

- Make sure your PC is connected to the Internet.
- Download the Raspberry Pi Pico W MicroPython UF2 file to a folder (e.g., Downloads) on your PC from the following link. At the time of writing this book the file had the name: **rp2-pico-w-20220909-unstable-v1.19.1-389-g4903e48e3.uf2**.

https://www.raspberrypi.com/documentation/microcontrollers/micropython.html#drag-and-drop-micropython

- Push and hold down the **BOOTSEL** button on your Pico.
- Connect your Pico to the USB port of your PC using a micro-USB cable while holding down the button.
- Wait a few seconds and let go of the **BOOTSEL** button.
- You should see the Pico appear as a removable drive with the name **RPI-RP2** as shown in Figure 2.1 (drive **E:** in this case).

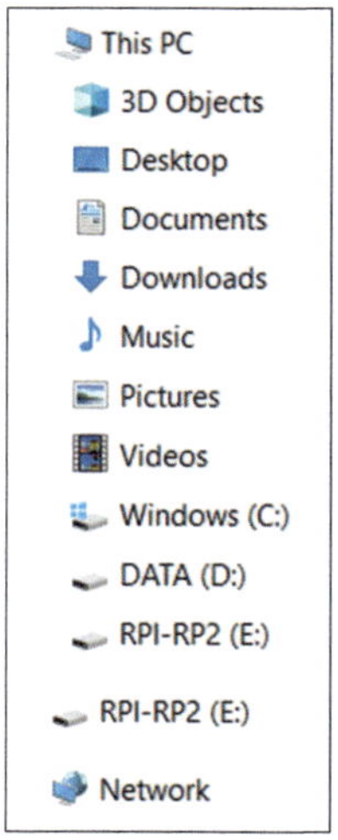

Figure 2.1: Pico as a removable drive RPI-RP2.

- Drag and drop the downloaded MicroPython UF2 file onto the RPI-RP2 volume. Your Pico will reboot, and you are now running MicroPython on your Pico.
- Powering down the Pico will not remove **MicroPython** from its memory.

2.3 Using the Thonny text editor from the PC

In this section, you will be learning how to use the Thonny on the PC to develop and run your programs.

First, you must install Thonny on your PC (if it is not already installed). The steps are:

- Go to the Thonny.org web site:

  https://thonny.org/

- Click on the link at the top right-hand side of the screen to install Thonny (see Figure 2.2).

Figure 2.2: Click to install Thonny.

- You should see an icon on the Desktop (Figure 2.3) of your PC. Double click to start Thonny.

Figure 2.3: Thonny icon on the Desktop.

- The startup screen of Thonny on your PC is shown in Figure 2.4.

Figure 2.4: Thonny startup screen on the PC.

- Click label **Python** at the bottom right-hand corner of the screen and click to select **MicroPython (Raspberry Pi Pico).**
- You are now ready to write your programs. Let us write a very simple Python statement to test the installation of MicroPython on your Pico.
- Enter the following statement at the lower part of the screen (in **Shell**):

print("hello from Thonny on PC")

- You should the message **hello from Thonny on PC** is displayed as shown in Figure 2.5.

```
Type "help()" for more information.
>>>
>>> print("hello from Thonny on PC")
 hello from Thonny on PC
>>> |
```

Figure 2.5: Displaying the message.

In this book, you will be using Thonny on the PC to write programs and to execute them on the Raspberry Pi Pico W.

Simple software-only example programs are given in the remaining sections of this chapter. The aim here has been to review the basic Python programming concepts. This book does not aim to teach the Python programming language. Interested readers can find many books and tutorials on the Internet for learning the Python programming language.

2.4 Writing a program using Thonny

In the previous section, you have learned how to execute a statement online using the Thonny **Shell**. Almost in all applications, you must write programs. As an example, the steps to write and run a very simple one-line program to display the message **Hello from program...** are given below:

- Enter the program statements at the upper part of the screen as shown in Figure 2.6.

Figure 2.6: Write the program on the upper part of the screen.

- Click **File** followed by **Save As** and give a name to your program. e.g., **FirstProg.** You have the option of storing the program either on your PC (**This computer**) or on the Pico (**Raspberry Pi Pico**). Click **Raspberry Pi Pico** to save it on the Pico (Figure 2.7). Enter the name of your program (**FirstProg**) and click **OK** (notice that the file is saved with the extension **.py**).

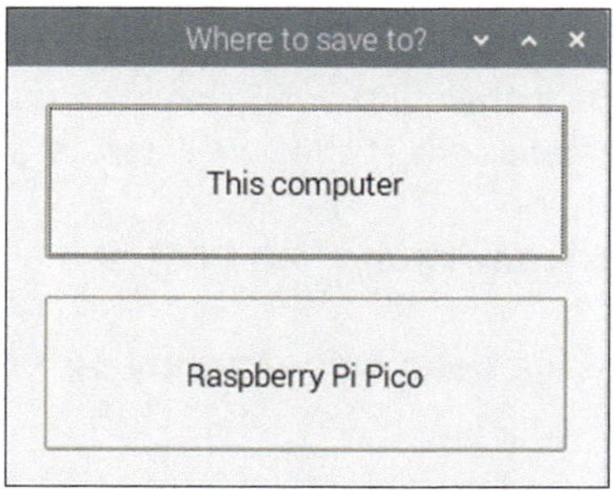

Figure 2.7: Click **Raspberry Pi Pico** *to save your program.*

- Click the green arrow icon at the top of the screen to run your program. The output of the program will be displayed in the lower **Shell** part of the screen as shown in Figure 2.8.

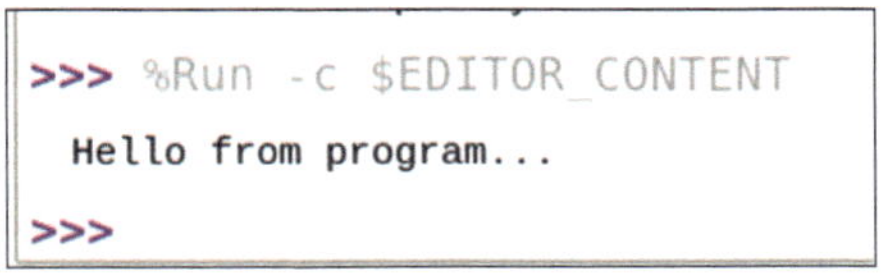

Figure 2.8: Output of the program.

2.5 Software only MicroPython programs using the Raspberry Pi Pico W

Example 1: Average of two numbers read from the keyboard

In this example, two numbers are read from the keyboard and their average is displayed. The aim of this example is to show how data can be read from the keyboard.

Solution 1

The program is named **Average** and the program listing and an example run of the program are shown in Figure 2.9. Function **input** is used to read the numbers in the form of strings from the keyboard. These strings are then converted into floating-point numbers and stored in variables **n1** and **n2**. The average is calculated by adding and then dividing the numbers by two. The result is displayed on the screen.

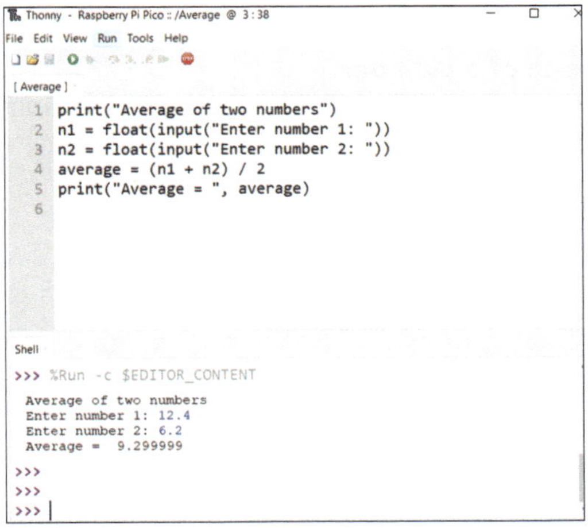

Figure 2.9: Program: ***Average*** *and sample run.*

Example 2: Average of 10 numbers read from the keyboard

In this example, 10 numbers are read from the keyboard and their average is displayed. The aim of this example is to show how a loop can be constructed in Python.

Solution 2

The program is named **Average10,** and the program listing and an example run of the program are shown in Figure 2.10. In this program, a loop is constructed which runs from 0 to 9 (i.e., 10 times). Inside this loop, the numbers are read from the keyboard, added to each other, and stored in a variable **sum**. The average is then calculated and displayed by dividing the **sum** by 10. Notice that a new-line is not printed after the print statements since the option **end = ' '** is used inside the print statement.

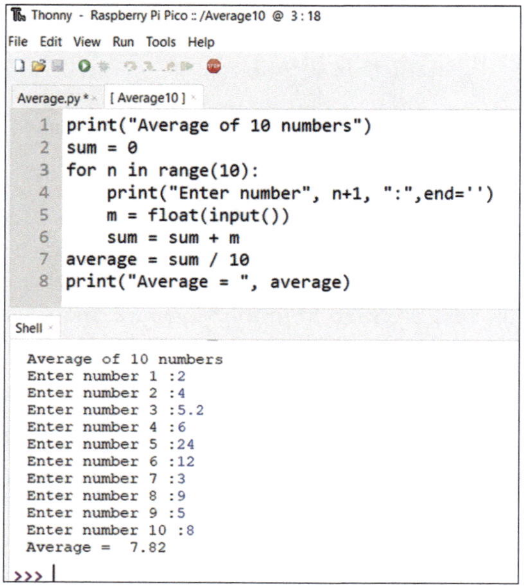

Figure 2.10: Program: **Average10** *and sample run.*

Example 3: Surface area of a cylinder

In this example, the radius and height of a cylinder are read from the keyboard and its surface area is displayed on the screen.

Solution 3

The program is named **CylArea,** and the program listing and an example run of the program are shown in Figure 2.11. The surface area of a cylinder is given by:

Surface area = $2\pi rh$

Where **r** and **h** are the radius and height of the cylinder respectively. In this program the **math** library is imported so that function **Pi** can be used in the program. The surface area of the cylinder is displayed after reading its radius and height.

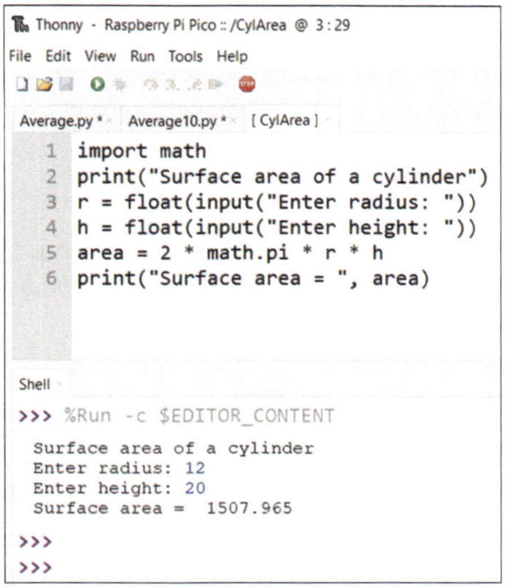

Figure 2.11: Program: ***CylArea*** *and sample run.*

Example 4: ºC to ºF conversion

In this example, the program reads Degrees Celsius from the keyboard and converts and displays the equivalent Degrees Fahrenheit.

Solution 4

The program is named **CtoF,** and the program listing and an example run of the program are shown in Figure 2.12. The formula to convert ºC to ºF is:

$$F = 1.8 \times C + 32$$

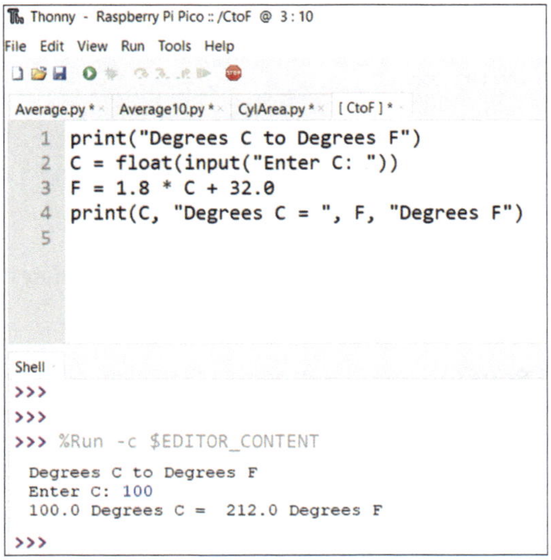

Figure 2.12: Program: ***CtoF*** *and sample run.*

Example 5: Surface area and volume of a cylinder – user function

In this example, the surface area and volume of a cylinder are calculated whose radius and height are given. The program uses a function to calculate and return the surface area and the volume.

Solution 5

The program is named **CylAreaSurf,** and the program listing and an example run of the program are shown in Figure 2.13. The surface area and the volume of a cylinder are given by:

Surface area = $2\pi rh$
Volume = $\pi r^2 h$

Where **r** and **h** are the radius and height of the cylinder respectively. Function **Calc** is used to get the radius and height of the cylinder. The function returns the surface area and volume to the main program which are displayed on the screen.

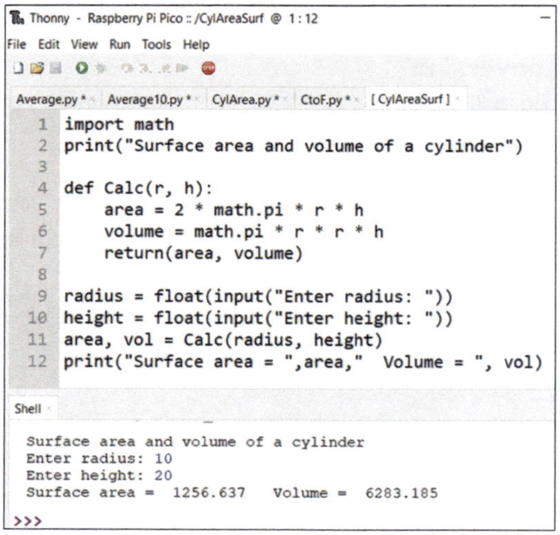

Figure 2.13: Program: ***CylAreaSurf*** *and sample run.*

Example 6: Table of squares of numbers

In this example, the squares of numbers from 1 to 11 are calculated and tabulated.

Solution 6

The program is named **Squares** and the program listing and an example run of the program are shown in Figure 2.14. Notice that **\t** prints a tab so that the data can be tabulated nicely.

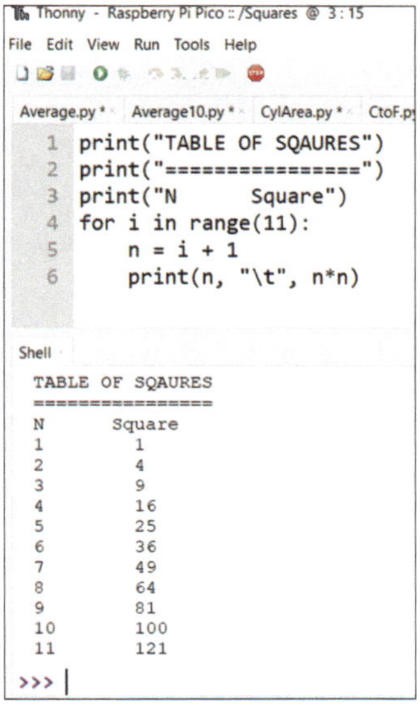

Figure 2.14: Program: ***Squares*** *and sample run.*

Example 7: Table of trigonometric sine

In this example, the trigonometric sine is tabulated from 0 to 45 degrees in steps of 5 degrees.

Solution 7

The program is named **Sines** and the program listing and an example run of the program are shown in Figure 2.15. It is important to notice that the arguments of the trigonometric functions must be in radians and not in degrees.

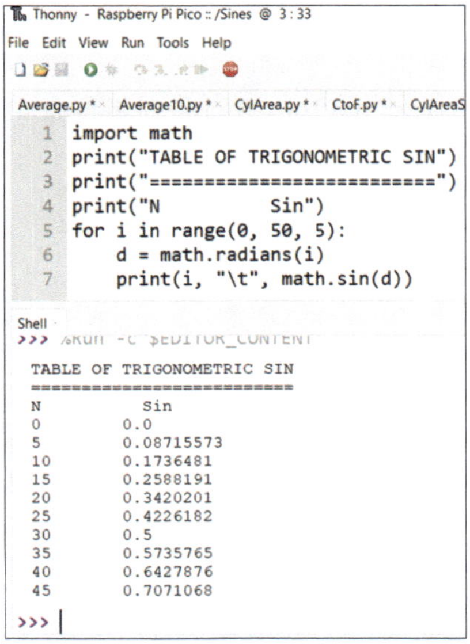

Figure 2.15: Program: ***Sines*** *and sample output.*

Example 8: Table of trigonometric sine, cosine, and tangent

In this example, the trigonometric sine, cosine, and tangent are tabulated from 0 to 45 degrees in steps of 5 degrees.

Solution 8

The program is named **Trig** and the program listing and an example run of the program are shown in Figure 2.16.

Trig.py

```
import math
print("          TABLE OF TRIGONOMETRIC FUNCTIONS")
print("          ================================")
print("N\t    Sin\t\t   Cos\t\t   Tan")
for i in range(0, 50, 5):
    d = math.radians(i)
    s = math.sin(d)
    c = math.cos(d)
    t = math.tan(d)
    print(i, "\t%9.7f\t%9.7f\t%9.7f" %(s,c,t))
```

Shell

```
          TABLE OF TRIGONOMETRIC FUNCTIONS
          ================================
N           Sin              Cos              Tan
0         0.0000000        1.0000000        0.0000000
5         0.0871557        0.9961946        0.0874887
10        0.1736481        0.9848078        0.1763270
15        0.2588191        0.9659258        0.2679492
20        0.3420201        0.9396926        0.3639702
25        0.4226182        0.9063078        0.4663076
30        0.5000000        0.8660254        0.5773502
35        0.5735765        0.8191521        0.7002075
40        0.6427876        0.7660444        0.8390996
45        0.7071068        0.7071068        1.0000000
```

Figure 2.16: Program: ***Trig*** *and sample output.*

Example 9: Trigonometric function of a required angle

In this example, an angle is read from the keyboard. Also, the user specifies whether the sine (s), cosine (c), or the tangent (t) of the angle is required.

Solution 9

The program is named **TrigUser,** and the program listing and an example run of the program are shown in Figure 2.17.

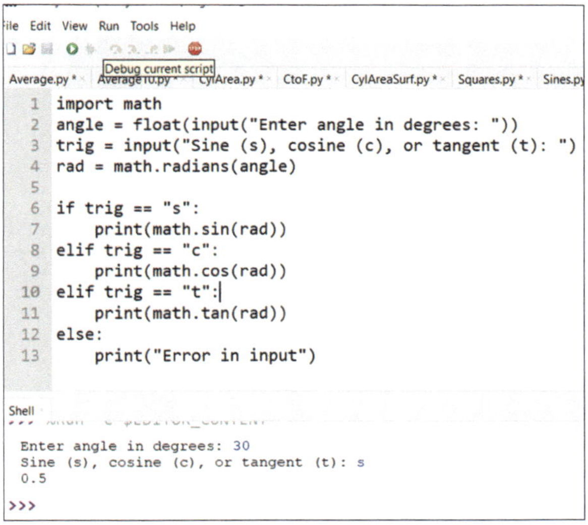

Figure 2.17: Program: ***TrigUser*** *and sample output.*

Example 10: Series and parallel resistors

This program calculates the total resistance of several series or parallel connected resistors. The user specifies whether the connection is in series or in parallel. Additionally, the number of resistors used is also specified at the beginning of the program.

Solution 10

When several resistors are in series then the resultant resistance is the sum of the resistance of each resistor. When the resistors are in parallel then the reciprocal of the resultant resistance is equal to the sum of the reciprocal resistances of each resistor.

Figure 2.18 shows the program listing (program: **Serpal**). At the beginning of the program a heading is displayed, and the program enters a **while** loop. Inside his loop the user is prompted to enter the number of resistors in the circuit and whether they are connected in series or in parallel. Function **str** converts a number into its equivalent string. e.g., number 5 is converted into string "5". If the connection is serial (mode equals to **'s'**) then the value of each resistor is accepted from the keyboard and the resultant is calculated and displayed on the screen. If on the other hand, the connection is parallel (mode equals to **'p'**) then again, the value of each resistor is accepted from the keyboard and the reciprocal of the number is added to the total. When all the resistor values are entered, the resultant resistance is displayed on the screen.

```
print("RESISTORS IN SERIES OR PARALLEL")
print("===============================")
yn = "y"

while yn == 'y':
   N = int(input("\nHow many resistors are there?: "))
   mode = input("Are the resistors series (s) or parallel (p)?: ")
```

```
    mode = mode.lower()
#
# Read the resistor values and calculate the total
#
    resistor = 0.0

    if mode == 's':
      for n in range(0,N):
         s = "Enter resistor " + str(n+1) + " value in Ohms: "
         r = int(input(s))
         resistor = resistor + r
      print("Total resistance = %d Ohms" %(resistor))

    elif mode == 'p':
      for n in range(0,N):
        s = "Enter resistor " + str(n+1) + " value in Ohms: "
        r = float(input(s))
        resistor = resistor + 1 / r
      print("Total resistance = %.2f Ohms" %(1 / resistor))
#
# Check if the user wants to exit
#
    yn = input("\nDo you want to continue?: ")
    yn = yn.lower()
```

Figure 2.18: Program: **Serpal***.*

Figure 2.19 shows a typical run of the program.

```
RESISTORS IN SERIES OR PARALLEL
===============================

How many resistors are there?: 3
Are the resistors series (s) or parallel (p)?: s
Enter resistor 1 value in Ohms: 100
Enter resistor 2 value in Ohms: 150
Enter resistor 3 value in Ohms: 200
Total resistance = 450 Ohms

Do you want to continue?: y

How many resistors are there?: 2
Are the resistors series (s) or parallel (p)?: p
Enter resistor 1 value in Ohms: 100
Enter resistor 2 value in Ohms: 100
Total resistance = 50.00 Ohms

Do you want to continue?: n
```

Figure 2.19: Typical run of the program.

Example 11: Words in reverse order

Write a program to read a word from the keyboard and then display the letters of this word in reverse order on the screen.

Solution 11

The required program listing is shown in Figure 2.20 (program: **Letters**). A word is read from the keyboard and stored in string variable **word**. Then the letters of this word are displayed in reverse order. An example run of the program is shown in Figure 2.20.

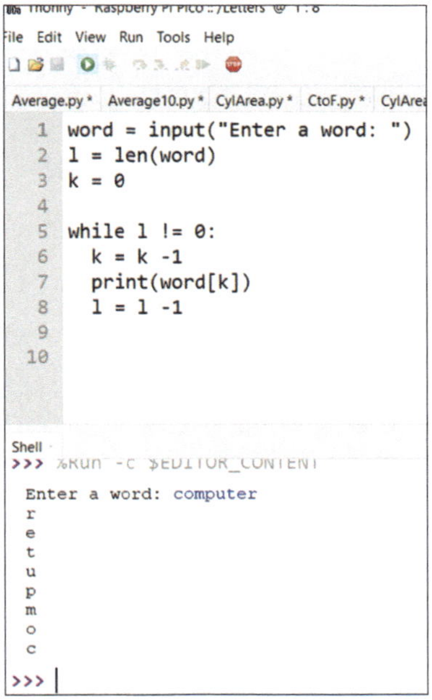

Figure 2.20: Program: ***Letters*** *and a sample output.*

Example 12: Calculator

Write a calculator program to carry out the four simple mathematical operations of addition, subtraction, multiplication, and division on two numbers received from the keyboard.

Solution 12

The required program listing is shown in Figure 2.21 (program: **Calc**). Two numbers are received from the keyboard and stored in variables **n1** and **n2**. Then, the required mathematical operation is received, and it is performed. The result, stored in variable **result**, is displayed on the screen. The user is given the option of terminating the program.

```
any = 'y'
while any == 'y':
   print("\nCalculator Program")
   print("==================")

   n1 = float(input("Enter first number: "))
   n2 = float(input("Enter second number: "))
   op = input("Enter operation (+-*/): ")

   if op =="+":
```

```
        result = n1 + n2
    elif op == "-":
        result = n1 - n2
    elif op == "*":
        result = n1 * n2
    elif op == "/":
        result = n1 / n2
    print("Result = %f" %(result))
    any = input("\nAny more (yn): ")
```

Figure 2.21: Program: ***Calc.***

An example run of the program is shown in Figure 2.22.

```
Calculator Program
==================
Enter first number: 25
Enter second number: 2
Enter operation (+-*/): *
Result = 50.000000

Any more (yn): y

Calculator Program
==================
Enter first number: 12
Enter second number: 2
Enter operation (+-*/): /
Result = 6.000000

Any more (yn): n

>>>
```

Figure 2.22: Example run of the program.

Example 13: File processing - writing

In this example, a text file called **MyFile.txt** will be created and text **Hello from Raspberry Pi Pico!** will be written to this file.

Solution 13

The program is named **Filew,** and its listing and an example run are shown in Figure 2.23. The file is opened in write (**w**) mode and the text is written in it using function **write**. Notice here that **fp** is the file handle.

```
print("Open the file and write the text")
fp = open("MyFile.txt", "w")
fp.write("Hello from Raspberry Pi Pico!\n")
fp.close()
print("End of file operation")
```

```
Shell
Open the file and write the text
End of file operation
```

Figure 2.23: Program: **Filew***.*

Example 14: File processing – reading

In this example, the text file **MyFile.txt** created in the previous example is opened and its contents is displayed on the screen.

Solution 14

The program is named **Filer,** and its listing and an example run are shown in Figure 2.24. The file is opened in read (**r**) mode and its contents is displayed.

```
print("Open the file and read its contents")
fp = open("MyFile.txt", "r")
str = fp.read(80)
fp.close()
print(str)
```

```
hell
Open the file and read its contents
Hello from Raspberry Pi Pico!
```

Figure 2.24: Program: **Filer***.*

Example 15: Squares and cubes of numbers

Write a program to tabulate the squares and cubes of numbers from 1 to 10.

Solution 15

The program is named **Cubes,** and its listing an example run are shown in Figure 2.25.

```
print("Squares and cubes of numbers")
print(' N  N*N  N*N*N')
for i in range(1,11):
    print('{0:2d} {1:3d} {2:4d}'.format(i, i*i, i*i*i))
```

Shell

```
Squares and cubes of numbers
 N  N*N  N*N*N
 1   1    1
 2   4    8
 3   9   27
 4  16   64
 5  25  125
 6  36  216
 7  49  343
 8  64  512
 9  81  729
10 100 1000
```

Figure 2.25: Program: ***Cubes*** *and example output.*

Example 16: Multiplication timetable

Write a program to read a number from the keyboard and then display the timetable for this number from 1 to 12.

Solution 16

The program is named **Times,** and its listing and an example run are shown in Figure 2.26.

```
num = int(input("Enter a number: "))
print("Timetable of number ", num)
for i in range(1, 13):
    print(num, 'x', i, '=', num*i)
```

Shell

```
Enter a number: 5
Timetable of number  5
5 x 1 = 5
5 x 2 = 10
5 x 3 = 15
5 x 4 = 20
5 x 5 = 25
5 x 6 = 30
5 x 7 = 35
5 x 8 = 40
5 x 9 = 45
5 x 10 = 50
5 x 11 = 55
5 x 12 = 60
```

Figure 2.26: Program: ***Times*** *and example output.*

Example 17: Odd or even

Write a program to read a number from the keyboard and check and display if this number is odd or even.

Solution 17

The program is named **OddEven,** and its listing and an example run are shown in Figure 2.27.

```
num = int(input("Enter a number: "))
if (num % 2) == 0:
   print("Number {0} is Even".format(num))
else:
   print("Number {0} is Odd".format(num))
```

```
Enter a number: 15
Number 15 is Odd
```

Figure 2.27: Program: ***OddEven*** *and example output.*

Example 18: Binary, octal, and hexadecimal

Write a program to read a decimal number from the keyboard. Convert this number into binary, octal, and hexadecimal and display on the screen.

Solution 18

The program is named **Conv,** and its listing and an example run are shown in Figure 2.28.

```
dec = int(input("Enter a number: "))
print("The decimal value of", dec, "is:")
print("in binary: ",bin(dec))
print("in octal: ",oct(dec))
print("in hexadecimal: ",hex(dec))
```

```
Enter a number: 254
The decimal value of 254 is:
in binary:  0b11111110
in octal:  0o376
in hexadecimal:  0xfe
```

Figure 2.28: Program: ***Conv*** *and example output.*

Example 19: Add two matrices

Write a program to add two given matrices and display the elements of the new matrix.

Solution 19

The program is named **AddMatrix,** and its listing and an example run are shown in Figure 2.29.

```
A = [[5,4,2],
     [6 ,3,1],
     [2 ,8,3]]

B = [[2,4,2],
     [0,4,10],
     [8,2,8]]

res = [[0,0,0],
       [0,0,0],
       [0,0,0]]

for i in range(len(A)):
   for j in range(len(A[0])):
       res[i][j] = A[i][j] + B[i][j]

for i in res:
   print(i)
```

```
>>> %Run -c $EDITOR_CONTENT

[7, 8, 4]
[6, 7, 11]
[10, 10, 11]
```

Figure 2.29: Program: **AddMatrix** *and example output.*

Example 20: Shapes

Write a program to use functions to calculate and display the areas of shapes: square, rectangle, triangle, circle, and cylinder. The sizes of the required sides should be received from the keyboard.

Solution 20

The areas of the shapes to be used in the program are as follows:

Square: side = a	area = a2
Rectangle: sides a, b	area = ab
Circle: radius r	area = пr2
Triangle: base b, height h	area = bh/2
Cylinder: radius r, height h	area = 2пrh

The required program listing is shown in Figure 2.30 (program: **areas**). A different function is used for each shape and the sizes of the sides are received inside the functions. The main program displays the calculated area for the chosen shape.

```
#------------------------------------------------------------
#              AREAS OF SHAPES
#              ===============
#
# This program calculates and displays the areas of various
# geometrical shapes of numbers in a list
#
# Author: Dogan Ibrahim
# File  : areas.py
# Date  : October, 2022
```

```
#------------------------------------------------------------
import math

def Square(a):                      # square
  return a * a

def Rectangle(a, b):                # rectangle
  return(a * b)

def Triangle(b, h):                 # triangle
  return(b * h / 2)

def Circle(r):                      # circle
  return(math.pi * r * r)

def Cylinder(r, h):                 # cylinder
  return(2 * math.pi * r * h)

print("AREAS OF SHAPES")
print("==============\n")
print("What is the shape?: ")

shape = input("Square (s)\nRectangle(r)\nCircle(c)\n\
Triangle(t)\nCylinder(y): ")

shape = shape.lower()
if shape == 's':
  a = float(input("Enter a side of the square: "))
  area = Square(a)
  s = "Square"
elif shape == 'r':
  a = float(input("Enter one side of the rectangle: "))
  b = float(input("Enter other side of the rectangle: "))
  area = Rectangle(a, b)
  s = "Rectangle"
elif shape == 'c':
  radius = float(input("Enter radius of the circle: "))
  area = Circle(radius)
  s = "Circle"
elif shape == 't':
  base = float(input("Enter base of the triangle: "))
  height = float(input("Enter height of the triangle: "))
  area = Triangle(base, height)
  s = "Triangle"
elif shape == 'y':
  radius = float(input("Enter radius of cylinder: "))
```

```
    height = float(input("Enter height of cylinder: "))
    area = Cylinder(radius, height)
    s = "Cylinder"

print("Area of %s is %f" %(s, area))
```

Figure 2.30: Program: ***areas.***

An example run of the program is shown in Figure 2.31.

```
Shell
AREAS OF SHAPES
===============

What is the shape?:
Square (s)
Rectangle(r)
Circle(c)
Triangle(t)
Cylinder(y): c
Enter radius of the circle: 10
Area of Circle is 314.159274
>>>
```

Figure 2.31: Example run of the program.

Example 21: Solution of a quadratic equation

Write a program to find the roots of a quadratic equation of the form:

$$ax^2 + bx + c = 0$$

Solution 21

The required program listing is shown in Figure 2.32 (program: **quadratic**). The coefficients a, b, and c are specified at the beginning of the program. The two roots are calculated and displayed on the screen.

```
[ quadratic ]
import cmath

a = 1
b = -5
c = 6

d = (b**2) - (4*a*c)            # the discriminant

root1 = (-b-cmath.sqrt(d))/(2*a)
root2 = (-b+cmath.sqrt(d))/(2*a)

print('The roots are {0} and {1}'.format(root1,root2))
```

```
Shell
The roots are (2+0j) and (3+0j)
```

Figure 2.32: Program: ***quadratic*** *and example output.*

Example 22: Matrix multiplication

Write a program to multiply two matrices.

Solution 22

The elements of the two matrices are specified at the beginning of the program. Figure 2.33 shows the program listing (program: **multmatrix**).

```
[ multmatrix ] ×
A = [[5,2,3],
     [2 ,1,3],
     [2 ,4,3]]

B = [[2,2,1,2],
     [3,2,3,0],
     [0,2,4,1]]

result = [[0,0,0,0],
          [0,0,0,0],
          [0,0,0,0]]

for i in range(len(A)):
   for j in range(len(B[0])):
       for k in range(len(B)):
           result[i][j] += A[i][k] * B[k][j]

for i in range(3):
  for j in range(4):
     print(result[i][j], end="")
     print(" ", end="")
  print("\n")
```

```
Shell
16 20 23 13

7 12 17 7

16 18 26 7
```

Figure 2.33: Program: ***multmatrix*** *and example output.*

Example 23: Factorial of a number

Write a program to calculate the factorial of a number entered from the keyboard.

Solution 23

The program listing is shown in Figure 2.34 (program: **factorial**).

```
[ factorial ]
numbr = int(input("Enter a number: "))
factorial = 1

# check if the number is negative, zero, or positive
if numbr < 0:
   print("Negative numbers do not have factorials")
elif numbr == 0:
   print("The factorial of 0 is 1")
else:
   for i in range(1, numbr + 1):
       factorial = factorial*i
   print("The factorial of",numbr,"is",factorial)
```

```
Shell
>>> %Run -c $EDITOR_CONTENT
 Enter a number: 5
 The factorial of 5 is 120
>>>
```

Figure 2.34: Program: ***factorial*** *and example output.*

Example 24: Compound interest

Write a program to calculate the compound interest given the initial value, interest rate, and the number of years.

Solution 24

Compound interest is calculated using the formula:

$$FV = IV(1 + i / 100) ** yr$$

$$\text{Compound interest} = FV - IV$$

Where, **FV** and **IV** are the future and initial values, **i** is the interest rate and **yr** is the number of years.

Figure 2.35 gives the program listing (program: **compound**).

```
[ compound ] ×
def compound_interest(initial, rate, years):

    FV = initial * (pow((1 + rate / 100), years))
    interest = FV - initial
    print("Compound interest is: ", interest)

IV = float(input("Enter initial value: "))
i = float(input("Enter interest rate :"))
yr = float(input("Enter number of years: "))

compound_interest(IV, i, yr)
```

```
Shell
>>> %Run -c $EDITOR_CONTENT
 Enter initial value: 1000
 Enter interest rate :10
 Enter number of years: 2
 Compound interest is:  210.0
>>>
```

Figure 2.35: Program: ***compound*** *and example output.*

Example 25: Guess the number

Write a program to generate a secret number between 1 and 50 and let the user guess this number in 5 attempts.

Solution 25

The program listing is shown in Figure 2.36 (program: **guess**).

```
[ guess ]
import random
total_guesses = 0

number = random.randint(1, 50)
print ("The secret number is between 1 and 50. You have 5 attempts")

while total_guesses < 5:
    guess = int(input("Your guess: "))
    total_guesses = total_guesses + 1

    if guess < number:
        print ("Too low...")
    if guess > number:
        print ("Too high...")
    if guess == number:
        break

if guess == number:
    print ("You guessed in {0} attempts".format(total_guesses))
else:
    print ("Sorry... The secret number was {0}".format(number))
```

```
Shell
The secret number is between 1 and 50. You have 5 attempts
Your guess: 30
Too high...
Your guess: 20
Too high...
Your guess: 15
You guessed in 3 attempts
```

Figure 2.36: Program: ***guess*** *and example output.*

Example 26: Numerical integration

Write a program to read a function from the keyboard and calculate and display the integral of this function between two given points.

Solution 26

The program listing is shown in Figure 2.37 (program: **integrate**). The integration is performed inside function **integrate**. This function is divided into equal small segments between the required lower and upper limits. The area of each segment is calculated. The total area is equal to the required integral of the function. The arguments of this function are the function to be integrated, low and high limits, and the number of points to consider.

```
def integrate(func, xlow, xhigh, n):
    length = xhigh - xlow
    sect = length / n
    area = 0
    for j in range(n):
        x = xlow + sect * j
        y = eval(func.replace("x", str(x)))
        secarea = y * sect
        area = area + abs(secarea)
    return (area)

function = input("Enter function: ")
lowl = float(input("Low limit:" ))
highl = float(input("High limit:"))
N = int(input("No of points: "))
```

```
r = integrate(function, lowl, highl, N)
print("integral of %s between %5.2f and %5.2f = %6.2f" %(function,lowl,highl,r))
```

Figure 2.37: Program: **integrate.**

Figure 2.38 shows an example run of the program where the following integral is calculated by using 100 segments and the result displayed on the Thonny screen:

$$\int_0^1 (x^2 + 2)dx$$

```
Enter function: x**2 + 2
Low limit:0
High limit:1
No of points: 100
integral of x**2 + 2 between  0.00 and  1.00 =   2.33
```

Figure 2.38: Example run of the program.

Example 27: Practise arithmetic

Write a program to display the following menu:

1. Addition
2. Subtraction
3. Multiplication
4. Division

Choice:

Generate two random integer numbers between 1 and 1000 and ask the user to carry out the required operation. Check the user's answer and display CORRECT or INCORRECT.

Solution 27

Figure 2.39 shows the required program (Program: **arithmetic**). Functions are used for the addition, subtraction, multiplication, and division. The number entered by the user is checked and appropriate message is displayed.

```
import random

def addition():
    n1 = random.randint(1, 1000)
    n2 = random.randint(1, 1000)
    print(n1, "+", n2, "= ")
    user_answer = int(input("Enter answer: "))
    correct_answer = n1 + n2
    return (user_answer, correct_answer)

def subtraction():
    n1 = random.randint(1, 1000)
```

```
    n2 = random.randint(1, 1000)
    print(n1, "-", n2, "= ")
    user_answer = int(input("Enter answer: "))
    correct_answer = n1 - n2
    return (user_answer, correct_answer)

def multiplication():
    n1 = random.randint(1, 1000)
    n2 = random.randint(1, 1000)
    print(n1, "X", n2, "= ")
    user_answer = int(input("Enter answer: "))
    correct_answer = n1 * n2
    return (user_answer, correct_answer)

def division():
    n1 = random.randint(1, 1000)
    n2 = random.randint(1, 1000)
    print(n1, "/", n2, "= ")
    user_answer = int(input("Enter answer: "))
    correct_answer = int(n1 / n2)
    return (user_answer, correct_answer)

def chk(user, correct):
    if user ==correct:
        print("CORRECT")
    else:
        print("INCORRECT")

print("   1. Addition")
print("   2. Subtraction")
print("   3. Multiplication")
print("   4. Integer Division")
ch = int(input("Choice: "))
if ch == 1:
    user,correct = addition()
elif ch == 2:
    user,correct = subtraction()
elif ch == 3:
    user,correct = multiplication()
else:
    user,correct = division()
chk(user,correct)
```

Figure 2.39: Program: **arithmetic**.

Figure 2.40 shows an example run of the program.

```
    1. Addition
    2. Subtraction
    3. Multiplication
    4. Division
Choice: 3
156 X 298 =
Enter answer: 46488
CORRECT
```

Figure 2.40: Example run.

Chapter 3 • Raspberry Pi Pico W LED Projects

3.1 Overview

In this chapter, you will be developing simple hardware projects with the Raspberry Pi Pico W, using the Thonny text editor. The following sub-headings will be given for each project where applicable:

- Title
- Description
- Aim
- Block diagram
- Circuit diagram
- Program listing
- Suggestions for future work

All the programs in this chapter have been developed using Thonny running on a PC, where the Raspberry Pi Pico W was connected to the USB port of the PC. A general-purpose bread-board was used to construct the projects.

3.2 Project 1: External flashing LED

Description: In this project, an external LED is connected to Pico, and the LED is flashed every second.

Aim: The aim of this project is to show how an external LED can be connected to Pico.

Block diagram: Figure 3.1 shows the block diagram of the project.

Figure 3.1: Block diagram of the project.

Circuit diagram: The circuit diagram of the project is shown in Figure 3.2 where the LED is connected to port pin GP1 of Pico through a 470-ohm current-limiting resistor.

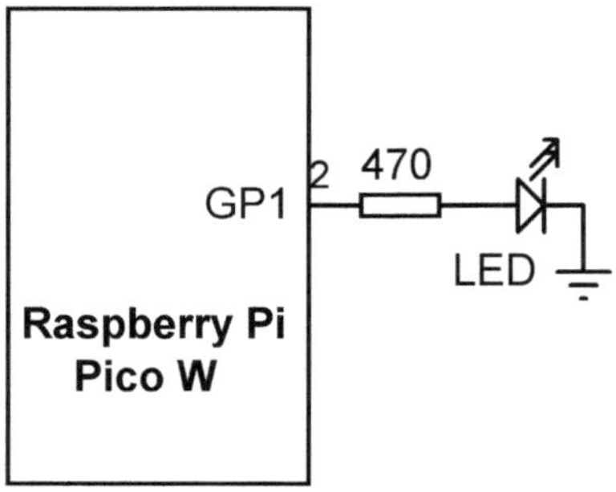

Figure 3.2: Circuit diagram of the project.

The LED can be connected in either current-sourcing or in current-sinking modes. In current-sourcing mode (Figure 3.3) the LED is turned ON when logic HIGH is applied to the port pin. In current-sinking mode (Figure 3.4) the LED is turned ON when logic LOW is applied to the port pin.

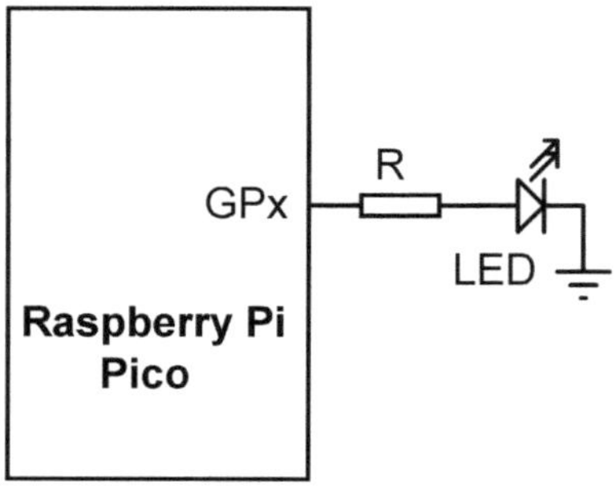

Figure 3.3: LED in current-sourcing mode.

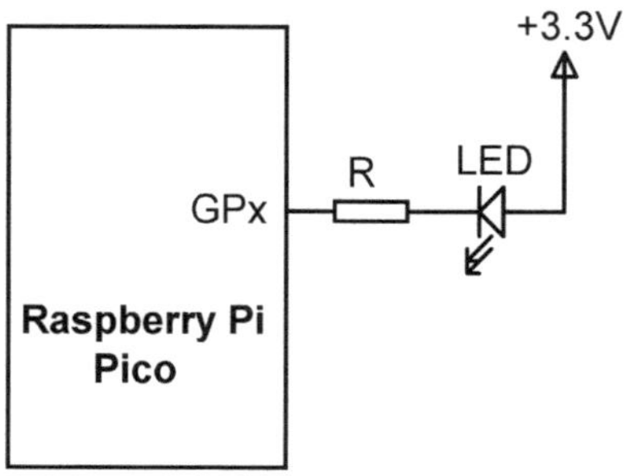

Figure 3.4: LED in current-sinking mode.

The required value of the current-limiting resistor can be calculated as follows:

In current-sourcing mode, assuming the output HIGH voltage is +3.3 V, voltage drop across the LED is 2 V, and the current through the LED is 3 mA, the required value of the current limiting resistor is:

R = (3.3 - 2) / 3 = 433 ohms, you will choose 470 ohms as the nearest resistor

Program listing: Figure 3.5 shows the program listing (Program: **ExtFlash.py**). At the beginning of the program LED at GP1 is configured as an output. A **while** loop is then formed which runs until stopped by the user. Inside this loop the LED is turned ON and OFF with one second delay between each output.

```
#----------------------------------------------------------
#                        FLASHING AN EXTERNAL LED
#                        ========================
#
# In this program an external LED is connected to port pin
# GP1 (pin 2). The LED is flashed every second
#
# Author: Dogan Ibrahim
# File  : ExtFlash.py
# Date  : October, 2021
#-----------------------------------------------------------
from machine import Pin
import utime

LED = Pin(1, Pin.OUT)              # LED at GP1

while True:                        # DO FOREVER
    LED.value(1)                   # LED ON
    utime.sleep(1)                 # Wait 1 second
    LED.value(0)                   # LED OFF
    utime.sleep(1)                 # Wait 1 second
```

Figure 3.5: Program: **ExtFlash.py**.

Figure 3.6 shows the Fritzing diagram of the project built on a breadboard (Raspberry Pi Pico is shown in this figure instead of the Pico W).

Figure 3.6: Project built on a breadboard.

The **machine** module supports the following functions (you will see in later chapters how to use these functions):

- Pin
- Timer
- ADC
- I^{2}C and SoftI2C
- SPI and SoftSPI
- WDT
- PWM
- UART

The general format of the machine.Pin function is (for more information, see link: https://docs.micropython.org/en/latest/library/machine.Pin.html):

```
machine.Pin(pin, mode, pull, value, alt)
```

The parameters of a pin can be re-initialized using the following function:

Pin.init(pin, mode, value, drive, at)

Where **pin** is the pin number.

Parameter **mode** can take the following values:

Pin.IN	- pin is configured as input
Pin.OUT	- pin is configured as output
Pin.OPEN_DRAIN	- pin is configured as open-drain output
Pin.ALT	- pin is configured as an alternative function

Parameter **pull** can take the following values:

NONE	- no internal pull-up or pull-down resistors
Pin.PULL_UP	- internal pull-up resistor enabled
Pin.PULL_DOWN	- internal pull-down resistor enabled

Parameter **value** is valid only for Pin.OUT and Pin.OPEN_DRAIN modes and it specifies the initial output pin value (if specified)

Parameter **alt** specifies an alternate function for the pin (port dependent). This parameter is valid only for PIN.ALT and Pin.ALT_OPEN_DRAIN modes.

A pin can be set/reset using one of the following functions:

Pin.value(1)	- set pin to logic 1
Pin.value(0)	- set pin to logic 0

Some other useful **machine** functions are:

machine.reset()	- reset the device (same as pressing the external RESET button)
machine.reset_cause)	- return the cause of the reset
machine.disable_irq()	- disable interrupt requests
machine.enable_irq()	- enable interrupt requests
machine.freq()	- returns the CPU frequency

3.3 Project 2: Flashing SOS signal

Description: In this project, an external LED flashes the SOS signal (three dots, followed by three dashes, followed by three dots) continuously. A dot is represented with the LED being ON for 0.25 seconds (Dot time) and a dash is represented with the LED being ON for 1 second (Dash time). The delay between the dots and dashes is set to 0.2 seconds (GAP time). This process is repeated continuously after 2 seconds of delay.

The block diagram and circuit diagram of this project are same as in Figure 3.1 and Figure 3.2 respectively.

Program listing. Figure 3.7 shows the program listing (Program: **SOS**). At the beginning of the program, the dot, dash, and gap times are defined. Then a loop is formed using a **while** statement. Inside this loop two **for** loops are formed, each iterating 3 times. The first loop displays three dots, while the second loop displays three dashes. This process is repeated after 2 seconds of delay.

```
#------------------------------------------------------------
#                      LED FLASHING SOS
#                      ================
#
# In this program the external LED at port GP1 flashes the
# SOS signal (... --- ...)
#
# Author: Dogan Ibrahim
# File  : SOS.py
# Date  : October, 2022
#-------------------------------------------------------------
from machine import Pin
import utime

Dot = 0.25                      # Dot time
Dash = 1.0                      # Dash time
```

```
Gap = 0.2                         # Gap time
ON = 1                            # ON
OFF = 0                           # OFF

LED = Pin(1, Pin.OUT)             # LED at GP1

while True:                       # DO FOREVER
    for i in range(0, 3):
        LED.value(ON)             # LED ON
        utime.sleep(Dot)          # Wait Dot time
        LED.value(OFF)            # LED OFF
        utime.sleep(Gap)          # Wait Gap time

    utime.sleep(0.5)              # 0.5 second delay

    for i in range(0, 3):
        LED.value(ON)             # LED ON
        utime.sleep(Dash)         # Wait Dash time
        LED.value(OFF)            # LED OFF
        utime.sleep(Gap)          # Wait Gap time

    utime.sleep(0.5)              # 0.5 second delay

    for i in range(0, 3):
        LED.value(ON)             # LED ON
        utime.sleep(Dot)          # Wait Dot time
        LED.value(OFF)            # LED OFF
        utime.sleep(Gap)          # Wait Gap time

    utime.sleep(2)                # Wait 2 seconds
```

Figure 3.7: Program: ***SOS****.*

Suggestions: You could easily replace the LED with a buzzer to make the SOS signal audible. There are two types of buzzers: active and passive. Passive buzzers require an audio signal to be sent to them and the frequency of the output signal depends on the frequency of the supplied signal. Active buzzers are ON/OFF type devices and they produce an audible sound when activated. In this project, you can use an active buzzer with a transistor switch (any NPN-type transistor can be used) as shown in Figure 3.8.

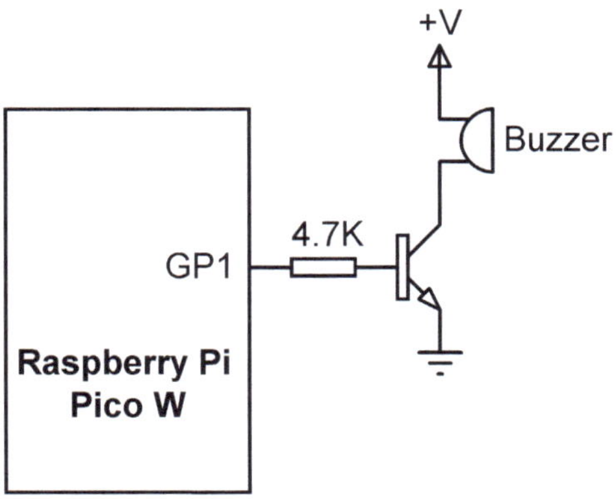

Figure 3.8: Using an active buzzer.

3.4 Project 3: Flashing LED – using a timer

Description: This project is very similar to Project 1 where an external LED is connected to port pin GP1 of the Pico. In this project, a timer is used to flash the LED every 500ms.

Aim: Timers are very important components of all microcontroller systems. The aim of this project s to show how a timer can be used in a program.

The block diagram and circuit diagram of this project are same as in Figure 3.1 and Figure 3.2 respectively.

Program listing. Figure 3.9 shows the program listing (Program: **LEDTimer**). The LED at port GP1 is configured as an output. Then, a timer is initialized which calls function **Flash_LED** twice (**frequency = 2.0 Hz, period = 0.5 seconds**) a second in a periodic manner. Notice that the LED is flashed using the **toggle** function.

```
#-----------------------------------------------------------
#              LED FLASHING USING A TIMER
#              ==========================
#
# In this program an external LED flashes twice a second
# using a timer
#
# Author: Dogan Ibrahim
# File  : LEDTimer.py
# Date  : October, 2022
#------------------------------------------------------------
from machine import Pin, Timer

LED = Pin(1, Pin.OUT)
tim = Timer()

def Flash_LED(timer):
```

```
    global LED
    LED.toggle()

tim.init(freq = 2.0, mode = Timer.PERIODIC, callback = Flash_LED)
```

Figure 3.9: Program: **LEDTimer**.

3.5 Project 4: Changing the LED flashing rate – using pushbutton interrupts

Description: In this project, an external LED is connected to port pin GP1 of the Pico as in the previous projects. Additionally, two pushbuttons are connected to port pins GP2 and GP3. At the start of the program, the LED flashes every second. Pushbutton at GP2 is named **Faster** and pressing this button flashes the LED faster. Similarly, pushbutton at GP3 is named **Slower**, and pressing this button flashes the LED slower.

Aim: The aim of this project is to show how pushbuttons can be connected to the Pico and how the state of a button can be read.

Block diagram: Figure 3.10 shows the block diagram of the project.

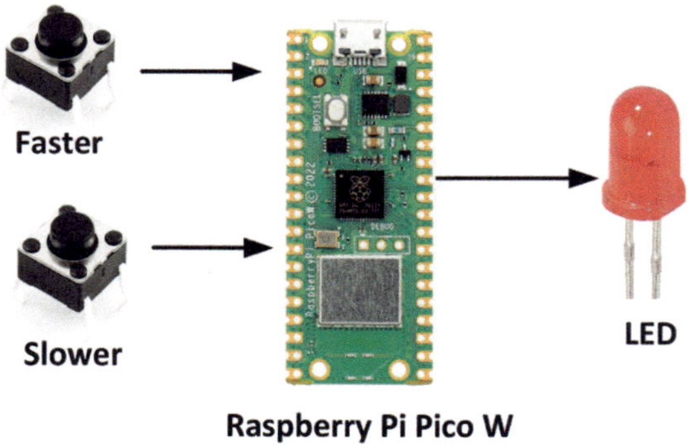

Figure 3.10: Block diagram of the project.

Circuit diagram: The circuit diagram of the project is shown in Figure 3.11. The LED is connected through a 470-ohm current-limiting resistor. The two pushbuttons are connected through 10K resistors. The default state of the pushbuttons is at logic 1 which is pulled up through the resistors. Pressing a pushbutton changes its output state to logic 0.

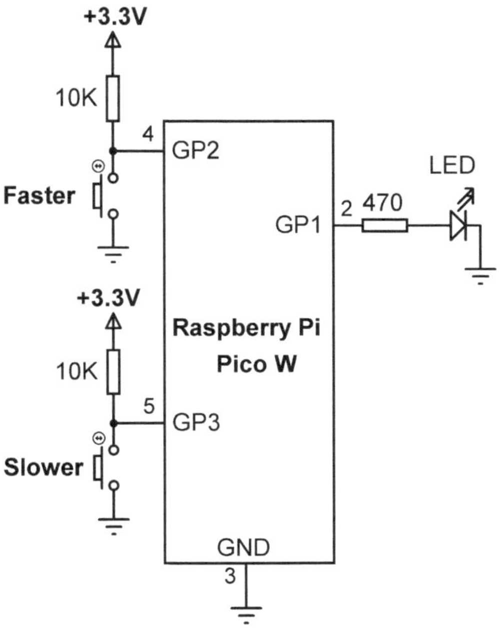

Figure 3.11: Circuit diagram of the project.

Program listing: Figure 3.12 shows the program listing (Program**: LEDrate**). At the beginning of the program LED is assigned to pin GP1, and pushbuttons **Faster** and **Slower** are assigned to ports GP2 and GP3 respectively. The default flashing rate is set to one second and is stored in variable **dly**. The program is external interrupt based. Pressing either of the pushbuttons creates an interrupt. For example, the interrupt for pushbutton **Faster** is configured using the following function call:

```
Faster.irq(handler=Flash_Faster, trigger=Faster.IRQ_FALLING)
```

Where **Flash_Faster** is the name of the interrupt service routine where **dly** is decremented. The interrupt is configured to happen on the falling edge of the pushbutton, i.e. when the pushbutton is pressed (the normal state of the pushbuttons is at logic 1, pulled-up by resistors). Inside the interrupt service routine **Flash_Faster** variable **dly** is declared as **global** so that it can be accessed. Its value is then decremented by 100 ms (0.1 seconds). Interrupt service routine for pushbutton **Slower** is done similarly where the delay is incremented by 100ms each time the button is pressed.

The other external interrupt modes are:

Pin.IRQ_FALLING	- interrupt on the falling edge (high-to-low)
Pin.IRQ_RISING	- interrupt on the rising edge (low-to-high)
Pin.IRQ_LOW_LEVEL	- interrupt on low level
Pin.IRQ_HIGH_LEVEL	- interrupt on high level

The above values can be OR'ed together to trigger on multiple events. You can also specify the interrupt priority with the keyword priority, where higher values represent higher priorities.

An interrupt **wake** parameter can be specified with the values **None**, **machine.IDLE**, **machine.SLEEP**, or **machine.DEEPSLEEP**.

Additionally, a parameter called **hard** with the values of **False** or **True** can be specified. If this parameter is set to **True**, then hardware interrupts are used which have a faster response.

```
#---------------------------------------------------------------
#              CHANGE THE LED FLASHING RATE
#              ============================
#
# In this program an external LED and two pushbuttons are used
# Pressing Faster flashes the LED faster, and pressing Slower
# flashes the LED slower
#
# Author: Dogan Ibrahim
# File  : LEDrate.py
# Date  : October, 2022
#---------------------------------------------------------------
from machine import Pin
import utime

LED = Pin(1, Pin.OUT)                       # LED at pin GP1
Faster = Pin(2, Pin.IN)                     # Faster at pin GP2
Slower = Pin(3, Pin.IN)                     # Slower at pin GP3
dly = 1.0                                   # Default delay

#
# This is the interrupt service routine. Whenever pushbutton
# Faster is pressed, the program jumps here and decrements
# delay to make the flashing faster
#
def Flash_Faster(Faster):
   global dly
   dly = dly - 0.1

#
# This is the interrupt service routine. Whenever pushbutton
# Slower is pressed, the program jumps here and increments
# delay to make the flashing slower
#
def Flash_Slower(Slower):
```

```
    global dly
    dly = dly + 0.1

#
# Configure the external interrupts
#
Faster.irq(handler=Flash_Faster,trigger=Faster.IRQ_FALLING)
Slower.irq(handler=Flash_Slower,trigger=Slower.IRQ_FALLING)

#
# Main program loop
#
while True:
    LED.value(0)                    # LED ON
    utime.sleep(dly)                # Delay dly
    LED.value(1)                    # LED OFF
    utime.sleep(dly)                # Delay dly
```

Figure 3.12: Program: **LEDrate***.*

Figure 3.13 shows Fritzing diagram of the project.

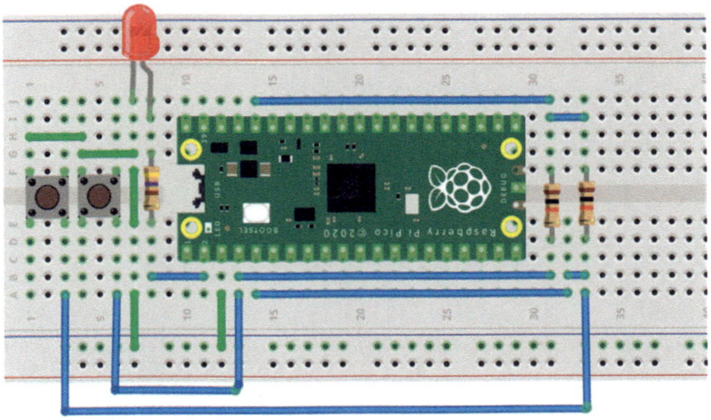

Figure 3.13: Project built on a breadboard.

Using the internal pull-up resistors

You can simplify the circuit diagram of Figure 3.11 by removing the external resistors and using the internal pull-up resistors of the Pico. The simplified circuit diagram is shown in Figure 3.14. The modified program listing (Program: **LEDrate2**) is shown in Figure 3.15 where the option **pull = Pin.PULL__UP** is added to the input configuration statements.

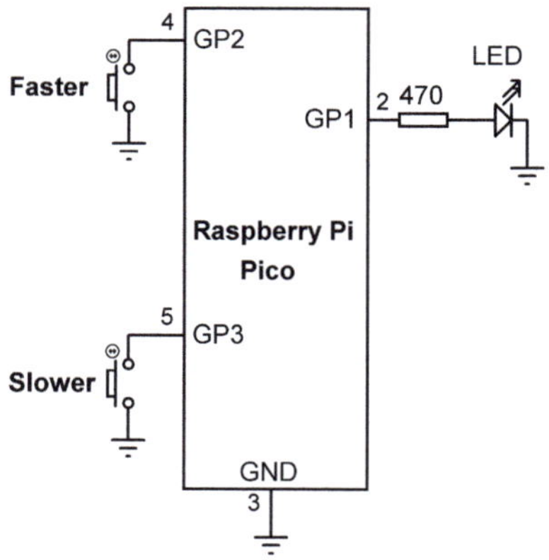

Figure 3.14: Modified circuit diagram.

```
#----------------------------------------------------------------
#              CHANGE THE LED FLASHING RATE
#              ============================
#
# In this program an external LED and two pushbuttons are used
# Pressing Faster flashes the LED faster, and pressing Slower
# flashes the LED slower. This program uses the internal pull-ups
#
# Author: Dogan Ibrahim
# File  : LEDrate2.py
# Date  : October, 2022
#-----------------------------------------------------------------
from machine import Pin
import utime

LED = Pin(1, Pin.OUT)                                # LED at pin GP1
Faster = Pin(2, Pin.IN, pull=Pin.PULL_UP)            # Faster at pin GP2
Slower = Pin(3, Pin.IN, pull=Pin.PULL_UP)            # Slower at pin GP3
dly = 1.0                                            # Default delay

#
# This is the interrupt service routine. Whenever pushbutton
# Faster is pressed, the program jumps here and decrements
# delay to make the flashing faster
#
def Flash_Faster(Faster):
   global dly
   dly = dly - 0.1

#
# This is the interrupt service routine. Whenever pushbutton
```

```
# Slower is pressed, the program jumps here and increments
# delay to make the flashing slower
#
def Flash_Slower(Slower):
    global dly
    dly = dly + 0.1

#
# Configure the external interrupts
#
Faster.irq(handler=Flash_Faster,trigger=Faster.IRQ_FALLING)
Slower.irq(handler=Flash_Slower,trigger=Slower.IRQ_FALLING)

#
# Main program loop
#
while True:
    LED.value(0)                # LED ON
    utime.sleep(dly)            # Delay dly
    LED.value(1)                # LED OFF
    utime.sleep(dly)            # Delay dly
```

Figure 3.15: Program: ***LEDrate2****.*

3.6 Project 5: Randomly flashing red, green, and blue LEDs – RGB

Description: In this project, an RGB LED is connected to Pico. The RGB flashes randomly, where the Red, Green, or the Blue LEDs are randomly ON or OFF.

Background Information: As shown in Figure 3.16, the RGB LED is a 4-pin device that incorporates Red, Green and Blue LEDs. Each colour LED is assigned a pin, where the fourth pin is the ground. By activating different LEDs at different brightness, you can generate many different colours. In this project, a common-cathode RGB LED is used. Notice that the cathode pin of the RGB LED is the longer pin.

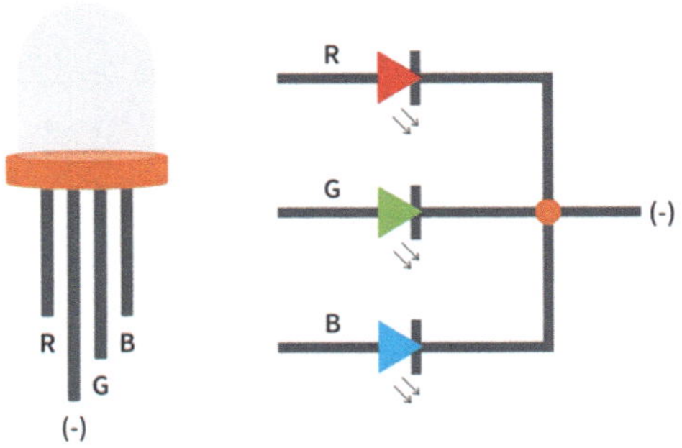

Figure 3.16: RGB LED (from CircuitBread).

Block Diagram: Figure 3.17 shows the block diagram of the project.

Figure 3.17: Block diagram of the project.

Circuit Diagram: The circuit diagram of the project is shown in Figure 3.18. The Red Green, and Blue pins are connected to port pins GP1, GP2, and GP3 respectively through 470-ohm current-limiting resistors.

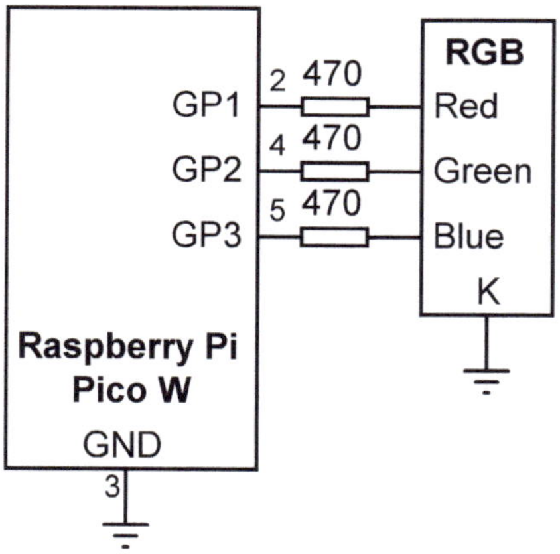

Figure 3.18: Circuit diagram of the project.

Program listing: Figure 3.19 shows the program listing (Program: **RGB**). Random numbers are generated either as 0 or 1 for each colour, and these numbers are used either to turn ON or OFF a colour LED.

```
#-----------------------------------------------------------
#                              RANDOMLY FLASHING RGB LED
#                              =========================
#
# In this program an RGB LED is connected to Pico.The three
# colours of the LED are flashed randomly every 500ms
#
# Author: Dogan Ibrahim
# File  : RGB.py
# Date  : October, 2022
#------------------------------------------------------------
from machine import Pin
import utime
import random

Red = Pin(1, machine.Pin.OUT)
Green = Pin(2, Pin.OUT)
Blue = Pin(3, Pin.OUT)

while True:
    r = random.randint(0, 1)
    g = random.randint(0, 1)
    b = random.randint(0, 1)
    Red.value(r)
    utime.sleep(0.2)
    Green.value(g)
    utime.sleep(0.2)
    Blue.value(b)
    utime.sleep(0.2)
```

Figure 3.19: Program: ***RGB****.*

3.7 Project 6: Binary counting LEDs

Description: In this project 8 LEDs are connected to Pico. The LEDs count up in binary from 0 to 255 as shown in Figure 3.20, with a one-second delay between each count.

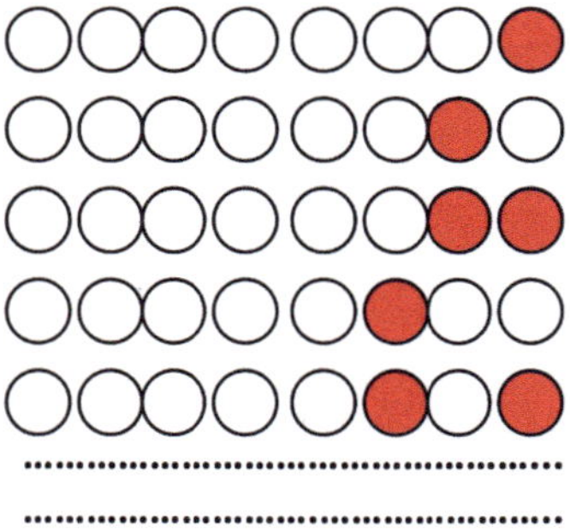

Figure 3.20: Binary counting LEDs.

Aim: The aim of this project is to show how a group of port pins can be combined and accessed as a parallel port.

Block diagram: Figure 3.21 shows the block diagram of the project.

Figure 3.21: Block diagram of the project.

Circuit diagram: The circuit diagram of the project is shown in Figure 3.22. The LEDs are connected to the Pico through 470-ohm current-limiting resistors.

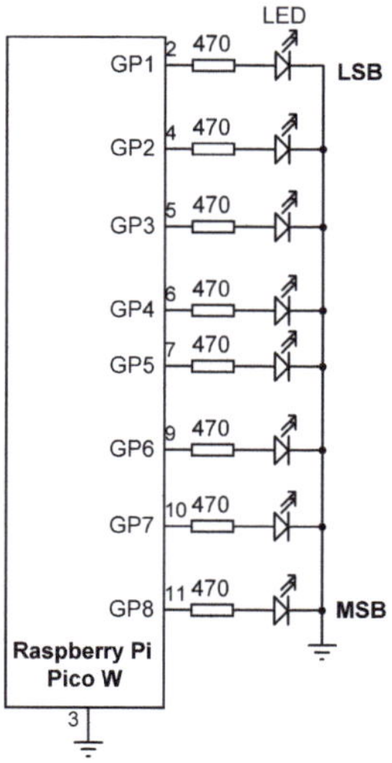

Figure 3.22: Circuit diagram of the project.

Program listing. Figure 3.23 shows the program listing (Program: **LEDCount**). All the 8 GPIO ports used in the project are configured as outputs using function **Configure_Port**. Notice that the **Configure_Port** function is general and list **DIR** sets the directions of the GPIO pins. An "O" sets as an output and an "I" sets as an input. Then, a loop is formed to execute forever and inside this loop, the LEDs count up by one in binary. Variable **cnt** is used as the counter. Function **Port_Output** is used to control the LEDs. This function can take integer numbers from 0 to 255 and it converts the input number (x) into binary using the built-in function **bin**. Then the leading "0b" characters are removed from the output string **b** (**bin** function inserts characters "0b" to the beginning of the converted string). Then, the converted string **b** is made up of 8 characters by inserting leading 0s. The string is then sent to the PORT bit by bit, starting from the most-significant bit position

```
#----------------------------------------------------------
#               BINARY COUNTING 8 LEDs
#               ======================
#
# In this program 8 LEDs are connected to Pico. The LEDs
# count up in binary every second
#
# Author: Dogan Ibrahim
# File  : LEDCount.py
# Date  : October, 2022
#-----------------------------------------------------------
from machine import Pin
import utime

PORT = [8, 7, 6, 5, 4,  3,  2,  1]             # port connections
DIR = ["0","0","0","0","0","0","0","0"]        # port directons
L = [0]*8

#
# This function configures the port pins as outputs ("0") or
# as inputs ("I")
#
def Configure_Port():
   for i in range(0, 8):
      if DIR[i] == "0":
         L[i] = Pin(PORT[i], Pin.OUT)
      else:
         L[i] = Pin(PORT[i], Pin.IN)
   return

#
# This function sends 8-bit data (0 to 255) to the PORT
#
def Port_Output(x):
```

```
    b = bin(x)                          # convert into binary
    b = b.replace("0b", "")             # remove leading "0b"
    diff = 8 - len(b)                   # find the length
    for i in range (0, diff):
       b = "0" + b                      # insert leading os

    for i in range (0, 8):
       if b[i] == "1":
          L[i].value(1)
       else:
          L[i].value(0)
    return

#
# Configure PORT to all outputs
#
Configure_Port()

#
# Main program loop. Count up in binary every second
#
cnt = 0
while True:
   Port_Output(cnt)                          # send cnt to port
   utime.sleep(1)                            # wait 1 second
   cnt = cnt + 1                             # increment cnt
   if cnt > 255:
      cnt = 0
```

Figure 3.23: Program: **LEDCount**.

3.8 Project 7: Lucky day of the week

Description: In this project, 7 LEDs are positioned in the form of a circle and are connected to the Raspberry Pi Pico. Each LED is assumed to represent a day of the week. Pressing a button generates a random number between 1 and 7 and lights up only one of the LEDs. The day name corresponding to this LED is assumed to be your lucky day of the week.

Block diagram: Figure 3.24 shows the block diagram of the project.

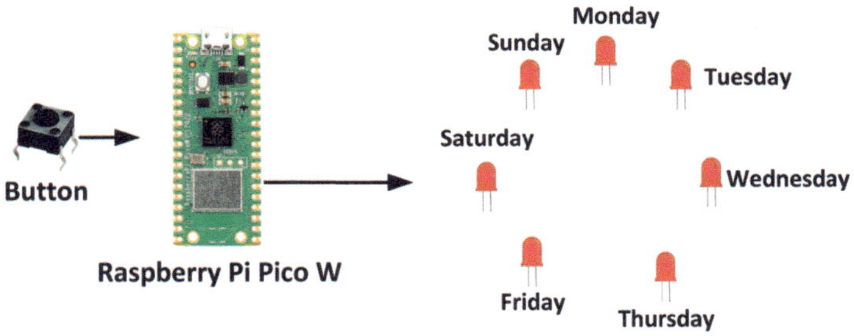

Figure 3.24: Block diagram of the project.

Circuit diagram: The circuit diagram of the project is shown in Figure 3.25, where 7 LEDs are connected to the Pico through current-limiting resistors. The button is connected to GP15. Normally the output of the button is at logic 1 and goes to logic 0 when the button is pressed.

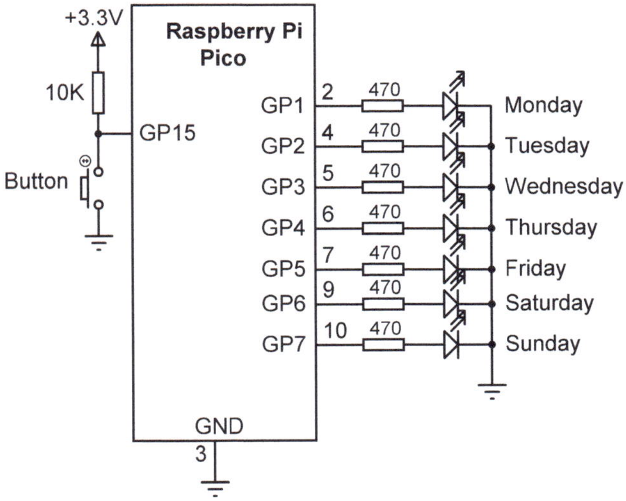

Figure 3.25: Circuit diagram of the project.

Program listing: Figure 3.26 shows the program listing (Program: **LuckyDay**). At the beginning of the program, all 8 LED GPIO pins are combined into a single port and are addressed as a single 8-bit port using the function **PORT_Output**. **utime.ticks_ms()** is used as the seed for the random number generator so that different sequences of numbers will be generated every time the program starts. This function returns an increasing milli-second counter with an arbitrary reference point, that wraps around after some value. An integer random number is generated between 1 and 7 and this number is used to turn ON one of the LEDs corresponding to a day of the week.

```
#-----------------------------------------------------------
#                       LUCKY DAY OF THE WEEK
#                       =====================
#
# In this program 7 LEDs are connected to Pico where each
# LED represents a day of the week. Pressing a button
# turns ON one of the LEDs randomly and this corresponds to
# your lucky day of the week
#
# Author: Dogan Ibrahim
# File  : LuckyDay.py
# Date  : October, 2022
#------------------------------------------------------------
from machine import Pin
import utime
import random

PORT = [7, 6, 5, 4, 3,  2,  1,  0]     # port connections
L = [0]*8
Button = Pin(15, Pin.IN)

#
# This function configures the LED ports as outputs
#
def Configure_Port():
   for i in range(0, 8):
         L[i] = Pin(PORT[i], Pin.OUT)

#
# This function sends 8-bit data (0 to 255) to the PORT
#
def Port_Output(x):
   b = bin(x)                          # convert into binary
   b = b.replace("0b", "")             # remove leading "0b"
   diff = 8 - len(b)                   # find the length
   for i in range (0, diff):
      b = "0" + b                      # insert leading os

   for i in range (0, 8):
      if b[i] == "1":
         L[i].value(1)
      else:
         L[i].value(0)
   return
#
# Configure PORT to all outputs
```

```
#
Configure_Port()

#
# Main program loop, check if Button is pressed
#
print("Press the Button to display your lucky number...")

random.seed(utime.ticks_ms())

while Button.value() == 1:                    # If Button not pressed
    pass
r = random.randint(1, 7)                      # Generate random number
r = pow(2, r-1)                               # LED to be turned ON
Port_Output(r)                                # Send to LEDs
```

Figure 3.26: Program: ***LuckyDay****.*

3.9 Project 8: Electronic dice

Description: In this project, 7 LEDs are arranged in the form of the faces of a dice and a pushbutton switch is used. When the button is pressed, the LEDs turn ON to display numbers 1 to 6 as if on a real dice. The display is turned OFF after 3 seconds, ready for the next game.

Aim: The aim of this project is to show how a dice can be constructed with 7 LEDs.

Block diagram: The block diagram of the project is shown in Figure 3.27.

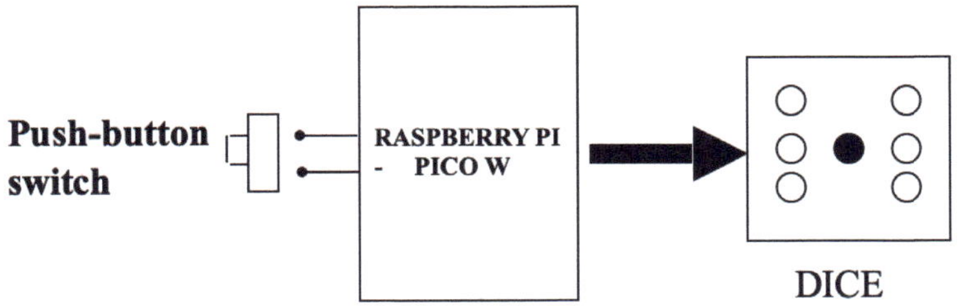

Figure 3.27: Block diagram of the project.

Figure 3.28 shows the LEDs that should be turned ON to display the 6 dice numbers.

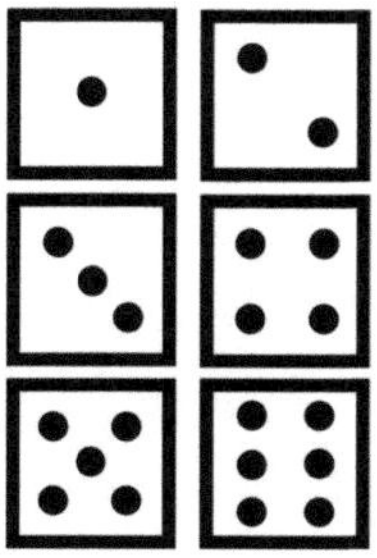

Figure 3.28: LED Dice numbers.

Circuit diagram: The circuit diagram of the project is shown in Figure 3.29. Here, 8 GPIO pins are collected to form a PORT. There are 7 LEDs, but 8 port pins are used in the form of a byte where the most-significant bit position is not used. The LEDs are mounted such that all the dice numbers can be displayed by turning ON the appropriate ones.

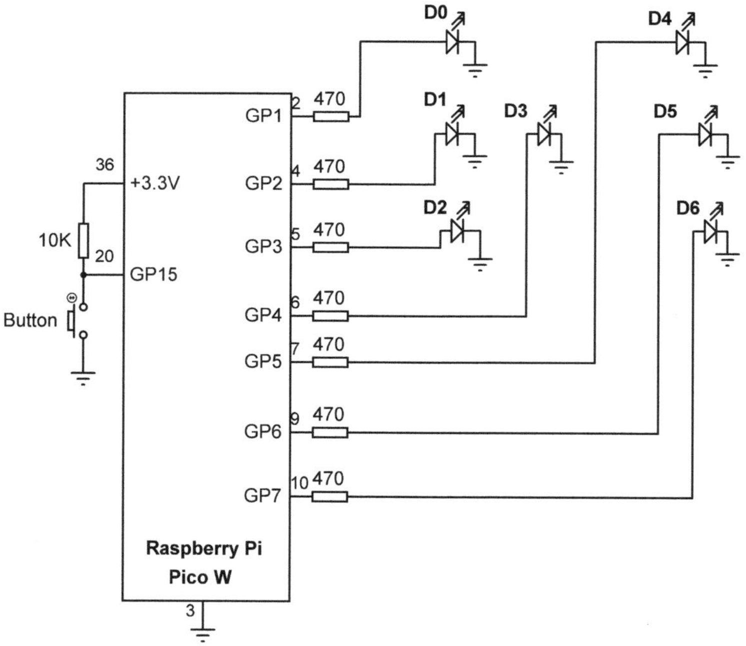

Figure 3.29: Circuit diagram of the project.

The pushbutton switch is connected to port pin GP15.

Table 3.1 gives the relationship between a dice number and the corresponding LEDs to be turned ON to imitate the faces of a real dice. For example, to display number 1 (i.e., only the middle LED is ON), you have to turn LED D3 ON. Similarly, to display number 4, you must turn ON D0, D2, D4 and D6.

Required number	LEDs to be turned on
1	D3
2	D0, D6
3	D0, D3, D6
4	D0, D2, D4, D6
5	D0, D2, D3, D4, D6
6	D0, D1, D2, D4, D5, D6

Table 3.1: Dice number and LEDs to be turned ON.

The relationship between the required number and the data to be sent to the PORT to turn on the correct LEDs is given in Table 3.2. For example, to display dice number 2, you must send hexadecimal 0x41 to the PORT. Similarly, to display number 5, you must send hexadecimal 0x5D to the PORT, and so on.

Required number	PORT data (Hex)
1	0x08
2	0x41
3	0x51
4	0x55
5	0x5D
6	0x77

Table 3.2: Required number and PORT data.

Program listing: The program is called **DICE** and the listing is shown in Figure 3.30. LED port pins are declared as a list in variable **PORT** and they are configured as outputs by using function **Configure_Port**. The bit pattern to be sent to the LEDs corresponding to each dice number is stored in hexadecimal format in a list called **DICE_NO** (see Table 3.2).

GP15 is configured as an input pin and the pushbutton switch is connected to this pin to simulate the "throwing" of a dice. The state of the pushbutton is checked in the main program and when the button is pressed, function DICE is called to display a dice number between 1 and 6 for 3 seconds. After this time all the LEDs are turned OFF to indicate that the program is ready to generate a new dice number. List **DICE_NO** is indexed to find the LEDs that should be turned ON, and the required bit pattern is sent to the PORT to display the dice number.

```
#-----------------------------------------------------------
#                             DICE PROGRAM
#                             ============
#
# In this program 7 LEDs are connected to Pico to simlate
# a dice. When a pushbutton is pressed the LEDs display a
# dice number between 1 and 6
#
# Author: Dogan Ibrahim
# File  : DICE.py
# Date  : October, 2022
#------------------------------------------------------------
from machine import Pin
import utime
import random

PORT = [8, 7, 6, 5, 4,  3,  2,  1]     # port connections
DICE_NO = [0, 0x08, 0x41, 0x51, 0x55, 0x5D, 0x77]
L = [0]*8
Button = Pin(15, Pin.IN)

#
# This function configures the LED ports as outputs
#
def Configure_Port():
   for i in range(0, 8):
         L[i] = Pin(PORT[i], Pin.OUT)

#
# This function sends 8-bit data (0 to 255) to the PORT
#
def Port_Output(x):
   b = bin(x)                            # convert into binary
   b = b.replace("0b", "")               # remove leading "0b"
   diff = 8 - len(b)                     # find the length
   for i in range (0, diff):
      b = "0" + b                        # insert leading os

   for i in range (0, 8):
      if b[i] == "1":
         L[i].value(1)
      else:
         L[i].value(0)
   return

#
```

```
# The program jumps here after the button is pressed
#
def DICE():
    n = random.randint(1, 6)                # generate a random number
    pattern = DICE_NO[n]                    # find the pattern
    Port_Output(pattern)                    # turn ON required LEDs
    utime.sleep(3)                          # wait for 3 seconds
    Port_Output(0)                          # turn OFF all LEDs
    return
#
# Configure PORT to all outputs
#
Configure_Port()

#
# Main program loop, check if Button is pressed
#
while True:
   if Button.value() == 0:                  # Button pressed?
      DICE()                                # Call DICE
   pass                                     # Do nothing
```

Figure 3.30: Program: **DICE**.

3.10 Project 9: Binary counter – Using the 74HC595 shift register

Description: In this project 8 LEDs are connected to the Raspberry Pi Pico W through a shift register and current-limiting resistors. The program counts up in binary through the LEDS as in Figure 3.20.

Background

The GPIO pin count of a microcontroller is usually limited, especially in complex control and monitoring projects. Shift registers are used to increase the pin capacity of a microcontroller. A shift register chip is controlled with only a few pins from a microcontroller and up to 8 outputs can be driven from the chip. It is also possible to cascade several shift registers and control many more outputs from a microcontroller. Shift registers receive serial data and then present this data on its output pins.

In this project, you will be using the popular 74HC595 shift register chip. Figure 3.31 shows the 74HC595 shift register chip. The pins of this chip are:

Pin	**Function**
1-7, 15	Q0-Q7 Parallel output pins
8	GND
9	Q7 serial output (for cascading)
10	MR active LOW master clear
11	SH_CP shift register clock pin

12	ST_CP (or latch) storage register clock pin
13	Active LOW output enable
14	DS serial data input
15	Vcc positive power supply

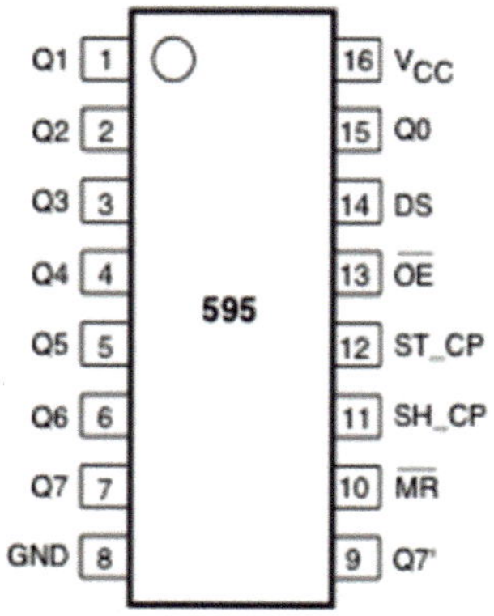

Figure 3.31: 74HC595 shift register chip.

Figure 3.32 shows the basic operation of the 74HC595 shift register. Serial data is clocked in through pin 14 with the help of the clock at pin 11 on the rising edge of the clock. Parallel output data is available at output pins 1-7 and 15 (pin 15 is the LSB) when the latch at pin 12 is raised to logic 1.

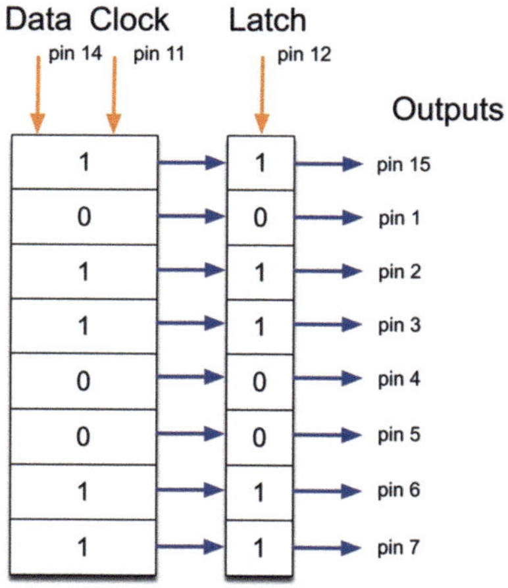

Figure 3.32: Operation of the 74HC595 shift register.

The operation of the 74HC595 shift register is shown by the PDL in figure 3.33.

```
BEGIN
      Configure DataPin (pin 14) as output
      Configure ClockPin (pin 11) as output
      Configure LatchPin (pin 12) as output
      Set clock = 0, latch = 0
      DO 8 times
             Set clock = 0
Send serial data bit in reverse order
Set clock = 1
      ENDDO
      Set latch = 1
END
```

Figure 3.33: PDL of the 74HC595 shift register.

Aim: The aim of this project is to show how a shift register can be used with the Raspberry Pi Pico W.

Block diagram: Figure 3.34 shows the block diagram of the project.

Figure 3.34: Block diagram of the project.

Circuit diagram: The circuit diagram of the project is shown in Figure 3.35. The LEDs are connected to the parallel outputs of the shift register through 470-ohm current-limiting resistors. The clock, latch, and serial data pins are connected to the Raspberry Pi Pico W pins. MR and OE pins of the shift register are not used in this project.

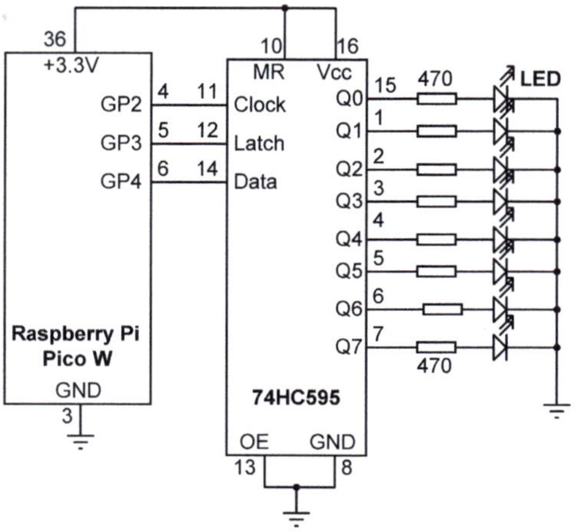

Figure 3.35: Circuit diagram of the project.

Program listing: Figure 3.36 shows the program listing (Program: **ShiftUpCnt**). At the beginning of the program Data, Clock, and Latch pins are configured as outputs. Function **ShiftOut** has 4 arguments. This function shifts out the data in serial form to the shift register. The parallel data is output at the Q0-Q7 pins when the latch pin of the shift register is raised. Inside the main program loop variable **Count** is incremented every second and the LEDs display binary counting up from 0 to 255. Notice that variable **Count** is converted into an 8-bit binary string inside the function **ShiftOut**.

```
-----------------------------------------------------------
#                              SHIFT REGISTER LED COUNTER
#                              ==========================
#
# In this program 8 LEDs are connected to a 74HC595 type
# shift register. The program controls the shift register to
# count up in binary every second using the LEDs
#
# Author: Dogan Ibrahim
# File  : ShiftUpCnt.py
# Date  : October, 2022
#------------------------------------------------------------
from machine import Pin
import utime

DataPin = Pin(4, Pin.OUT)
ClockPin = Pin(2, Pin.OUT)
LatchPin = Pin(3, Pin.OUT)

def ShiftOut(input, DS, CLK, Latch):
    CLK.value(0)
    Latch.value(0)
```

```
    for k in range(8):
        CLK.value(0)
        val = "{0:08b}".format(input)
        DS.value(int(val[k]))
        CLK.value(1)

    Latch.value(1)

Count = 0
while True:
    ShiftOut(Count, DataPin, ClockPin, LatchPin)
    Count = Count + 1
    if Count > 255:
        Count = 0
    utime.sleep(1)
```

Figure 3.36: Program: ***ShiftUpCnt****.*

Making the ShiftOut function a module

Because the shift register is likely to be used in many projects, it is worthwhile to make the function **ShiftOut** an importable module. The steps are:

- Create a new file and copy the **ShiftOut** code to this file
- Save the file as **ShiftOut.py** on your Raspberry Pi Pico W (male sure extension **.py** is included)
- Import the module at the beginning of your program
- Configure the data, clock, and latch pins as outputs

Figure 3.37 shows an example program (Program: **ShiftUpCnt2**) where module **ShiftOut** is imported to the program.

```
#----------------------------------------------------------
#                              SHIFT REGISTER LED COUNTER
#                              ==========================
#
# In this program 8 LEDs are connected to a 74HC595 type
# shift register. The program controls the shift register to
# count up in binary every second using the LEDs
# This program uses the ShiftOut module
#
# Author: Dogan Ibrahim
# File  : ShiftUpCnt2.py
# Date  : October, 2022
#-----------------------------------------------------------
from machine import Pin
```

```
import utime
from ShiftOut import ShiftOut

DataPin = Pin(4, Pin.OUT)
ClockPin = Pin(2, Pin.OUT)
LatchPin = Pin(3, Pin.OUT)

Count = 0
while True:
    ShiftOut(Count, DataPin, ClockPin, LatchPin)
    Count = Count + 1
    if Count > 255:
        Count = 0
    utime.sleep(1)
```

Figure 3.37: Importing module ShiftOut.

3.11 Project 10: Chasing LEDs - Using the 74HC595 shift register

Description: In this project, 8 LEDs are connected to the Raspberry Pi Pico W through a shift register and current-limiting resistors as in the previous project. The program turns ON/OFF the LEDs in a chasing manner every second as shown in Figure 3.38.

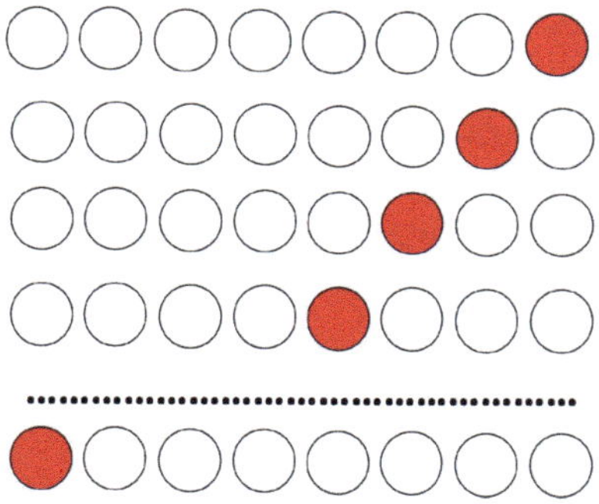

Figure 3.38: Chasing LEDs.

The block diagram and circuit diagram of the project are in Figure 3.34 and Figure 3.35 respectively.

Program listing: Now that you have the **ShiftOut** module, the program is very simple and consists of several lines as shown in Figure 3.39 (Program: **rotate**).

```
#-----------------------------------------------------------
#                              SHIFT REGISTER ROTATE LEDs
#                              ==========================
#
# In this program 8 LEDs are connected to a 74HC595 type
# shift register. The program rotates the LEDs as described
# in the text
#
# Author: Dogan Ibrahim
# File  : rotate.py
# Date  : October, 2022
#------------------------------------------------------------
from machine import Pin
import utime
from ShiftOut import ShiftOut

DataPin = Pin(4, Pin.OUT)
ClockPin = Pin(2, Pin.OUT)
LatchPin = Pin(3, Pin.OUT)

Count = 1
while True:
    ShiftOut(Count, DataPin, ClockPin, LatchPin)
    Count = Count << 1
    if Count > 128:
        Count = 1
    utime.sleep(1)
```

Figure 3.39: Program: **rotate**.

3.12 Project 11: Turning a selected LED ON - Using the 74HC595 shift register

Description: In this project, 8 LEDs are connected to the Raspberry Pi Pico W through a shift register and current-limiting resistors as in the previous project. The program turns ON a selected LED. The LEDs are numbered 1 to 8, 1 being the LED at the LSB position. For example, selecting 2 turns ON the second LED from the right-hand side.

The block diagram and circuit diagram of the project are in Figure 3.34 and Figure 3.35 respectively.

Program listing: Figure 3.40 shows the program listing (Program: **LEDselect**).

```
#------------------------------------------------------------
#               SHIFT REGISTER SELECT AN LEDs
#               ==============================
#
# In this program 8 LEDs are connected to a 74HC595 type
# shift register. The program turns ON a selected LED
#
# Author: Dogan Ibrahim
# File  : LEDselect.py
# Date  : October, 2022
#-------------------------------------------------------------
from machine import Pin
import utime
from ShiftOut import ShiftOut

DataPin = Pin(4, Pin.OUT)
ClockPin = Pin(2, Pin.OUT)
LatchPin = Pin(3, Pin.OUT)

while True:
    sel = int(input("Selected LED (1-8): "))
    sel = sel - 1
    r = 2**sel
    ShiftOut(r, DataPin, ClockPin, LatchPin)
    utime.sleep(3)
```

Figure 3.40: Program: ***LEDselect****.*

An example run of the program is shown in Figure 3.41.

```
Selected LED (1-8): 2
Selected LED (1-8): 4
Selected LED (1-8): 1
Selected LED (1-8): 7
Selected LED (1-8): 8
Selected LED (1-8): 7
Selected LED (1-8):
```

Figure 3.41: Example run.

3.13 Project 12: Randomly flashing LEDs – Using the 74HC595 shift register

Description: In this project, 8 LEDs are connected to the Raspberry Pi Pico W through a shift register and current-limiting resistors as in the previous project. The program flashes the LEDs randomly as in Christmas tree lights.

The block diagram and circuit diagram of the project are in Figure 3.34 and Figure 3.35 respectively.

Program listing: Figure 3.42 shows the program listing (Program: XMAS).

```
#-----------------------------------------------------------
#              SHIFT REGISTER RANDOM FLASHING LEDs
#              ===================================
#
# In this program 8 LEDs are connected to a 74HC595 type
# shift register. The program flashes the LEDs randomly
#
# Author: Dogan Ibrahim
# File  : XMAS.py
# Date  : October, 2022
#------------------------------------------------------------
from machine import Pin
import utime
import random
from ShiftOut import ShiftOut

DataPin = Pin(4, Pin.OUT)
ClockPin = Pin(2, Pin.OUT)
LatchPin = Pin(3, Pin.OUT)

while True:
    n = random.randint(1, 255)
    ShiftOut(n, DataPin, ClockPin, LatchPin)
    utime.sleep(0.2)
```

Figure 3.42: Program: ***XMAS****.*

3.14 Project 13: Traffic lights

Description: In this project, a program is written to control traffic lights at a simple junction. Red, yellow, and green LEDs are used to simulate the traffic lights.

It is assumed that traffic flows from the left-hand side of the road (i.e., the steering wheels of cars are on the right-hand side). In this project, the junction consists of two roads: NORTH ROAD and EAST ROAD. The traffic lights are located at the junction as shown in Figure 3.43.

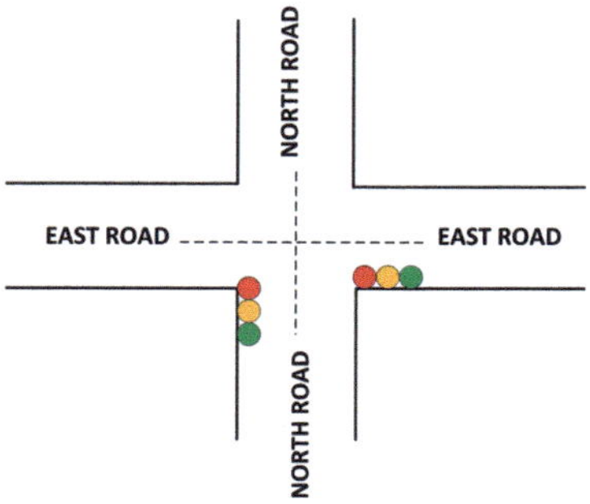

Figure 3.43: Example junction.

The traffic light sequence in the UK is as follows (Figure 3.44). Note: yellow colour is amber colour:

ROAD 1	**ROAD 2**
RED	GREEN
RED+YELOW	YELLOW
GREEN	RED
YELLOW	RED+YELLOW
RED	GREEN

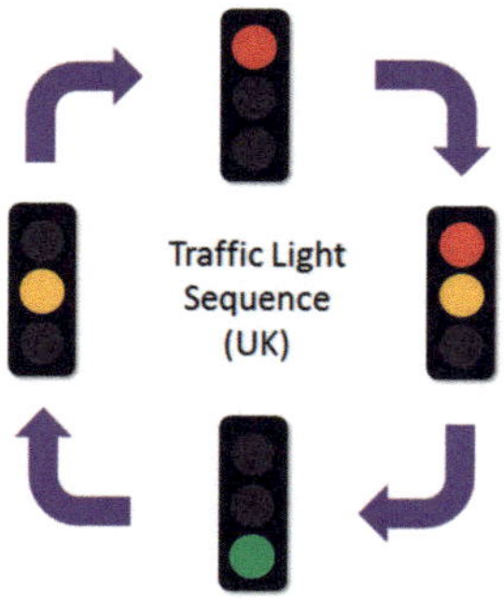

Figure 3.44: Traffic sequence in the UK.

The timings used in this project are shown in Table 3.3, where the green on EAST ROAD is for 10 seconds, and the green on NORTH road is for 4 seconds. The cycle time of the lights is 27 seconds, having the following values:

EAST ROAD	**NORTH ROAD**
10 s green	18 s red
12 s red	2 s red + yellow
2 s red + yellow	4 s green
3 s yellow	3 s yellow

Time (seconds)	EAST ROAD	NORTH ROAD
0	GREEN	RED
1	GREEN	RED
2	GREEN	RED
3	GREEN	RED
4	GREEN	RED
5	GREEN	RED
6	GREEN	RED
7	GREEN	RED
8	GREEN	RED
9	GREEN	RED
10	YELLOW	RED
11	YELLOW	RED
12	YELLOW	RED
13	RED	RED
14	RED	RED + AMBER
15	RED	RED + AMBER
16	RED	GREEN
17	RED	GREEN
18	RED	GREEN
19	RED	GREEN
20	RED	YELLOW
21	RED	YELLOW
22	RED	YELLOW
23	RED	RED
24	RED + AMBER	RED
25	RED + AMBER	RED
26	GREEN	RED
27	GREEN	RED

Table 3.3: Timing used in the project.

Block diagram: The block diagram of the project is shown in Figure 3.45. There are 6 LEDs used, 3 for each road.

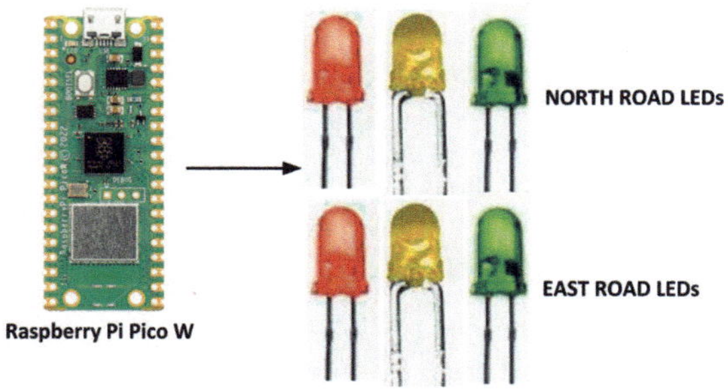

Figure 3.45: Block diagram of the project.

Circuit diagram: Figure 3.46 shows how the LEDs are connected to the Raspberry Pi Pico W. The red, yellow, and green LEDs of the NORTH ROAD are connected to ports GP10, GP11, and GP12 respectively. The red, yellow, and green LEDs of the EAST ROAD are connected to port pins GP13, GP14, and GP15 respectively.

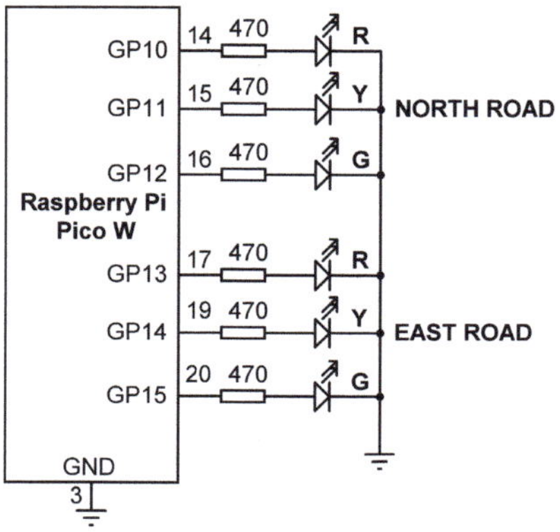

Figure 3.46: Circuit diagram of the project.

Construction: The project was constructed on a breadboard and the jumper wires used to connect the LEDs and resistors were connected to the development board.

Program listing: Figure 3.47 shows the program listing (Program: **traffic**). At the beginning of the program, the LEDs are assigned to output ports and all the LEDs are turned OFF. The main program runs forever in a **while** loop. Inside this loop, the LEDs at both roads are turned ON and OFF at the times specified by Table 3.3.

```
#------------------------------------------------------------
#                     TRAFFIC LIGHTS
#                     ==============
#
# This is a traffic lights project. A junction is simulated
# with the details given in the text
#
# Author: Dogan Ibrahim
```

```
# File   : traffic.py
# Date   : October 2022
#------------------------------------------------------------
from machine import Pin
import utime

#
# Configure LEDs
#
NorthR = Pin(10, Pin.OUT)                    # NORTH ROAD red
NorthY = Pin(11, Pin.OUT)                    # NORTH ROAD yellow
NorthG = Pin(12, Pin.OUT)                    # NORTH ROAD green
EastR  = Pin(13, Pin.OUT)                    # EAST ROAD red
EastY  = Pin(14, Pin.OUT)                    # EAST ROAD yellow
EastG  = Pin(15, Pin.OUT)                    # EAST ROAD green

#
# All LEDs OFF
#
NorthR.value(0)
NorthY.value(0)
NorthG.value(0)
EastR.value(0)
EastY.value(0)
EastG.value(0)

#
# Start of traffic sequence
#
while True:
    NorthG.value(1)
    EastR.value(1)
    utime.sleep(10)

    NorthG.value(0)
    NorthY.value(1)
    utime.sleep(3)

    NorthY.value(0)
    NorthR.value(1)
    utime.sleep(1)

    EastY.value(1)
    utime.sleep(2)

    EastR.value(0)
```

```
EastY.value(0)
EastG.value(1)
utime.sleep(4)

EastG.value(0)
EastY.value(1)
utime.sleep(3)

EastY.value(0)
EastR.value(1)
utime.sleep(1)

NorthY.value(1)
utime.sleep(2)

NorthY.value(0)
NorthR.value(0)
```

Figure 3.47: Program: **traffic**.

3.15 Project 14: Simple logic probe

Description: This is a simple logic probe project. A logic probe is used to indicate the logic state of an unknown digital signal. In a typical application, a test load (probe) is used to detect the unknown signal, and two different colour LEDs are used to indicate the logic status. In this project, the RED LED indicates logic 0 and the GREEN LED indicates logic 1.

Circuit diagram: Figure 3.48 shows the circuit diagram of the project. The unknown digital signal is applied to port pin GP13, RED and GREEN LEDs are connected to GP15 and GP14 respectively.

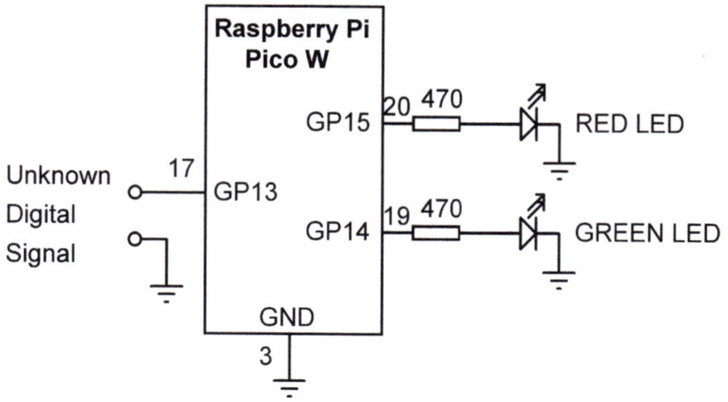

Figure 3.48: Circuit diagram of the project.

Program listing: Figure 3.49 shows the program listing (Program: **probe**).

```
#------------------------------------------------------------
#              LOGIC PROBE
#              ===========
#
# This is a logic probe program. The state (0 or 1) of an
# unknown digital signal is indicated on two LEDs
#
# Author: Dogan Ibrahim
# File  : probe.py
# Date  : October, 2022
#-------------------------------------------------------------
from machine import Pin

RED = Pin(15, Pin.OUT)                          # RED LED
GREEN = Pin(14, Pin.OUT)                        # GREEN LED
PROBE = Pin(13, Pin.IN)                         # PROBE

while True:
    if PROBE.value() == 1:                      # Logic 1?
        RED.value(0)                            # RED OFF
        GREEN.value(1)                          # GREEN ON
    else:                                       # Logic 0
        RED.value(1)                            # RED ON
        GREEN.value(0)                          # GREEN OFF
```

Figure 3.49: Program: **probe***.*

3.16 Project 15: Advanced logic probe

Description: The problem with this logic probe is that one of the LEDs is always ON even if the probe is not connected to any digital signal, or if the signal is at a high-impedance state (i.e., tristate). You can develop the project further so that the high-impedance state can also be detected and none of the LEDs are turned ON.

Circuit diagram: Figure 3.50 shows the modified circuit diagram. Notice here that a transistor (BC108 or any NPN-type transistor) is used at the front end of the circuit. The operation of the circuit is as follows:

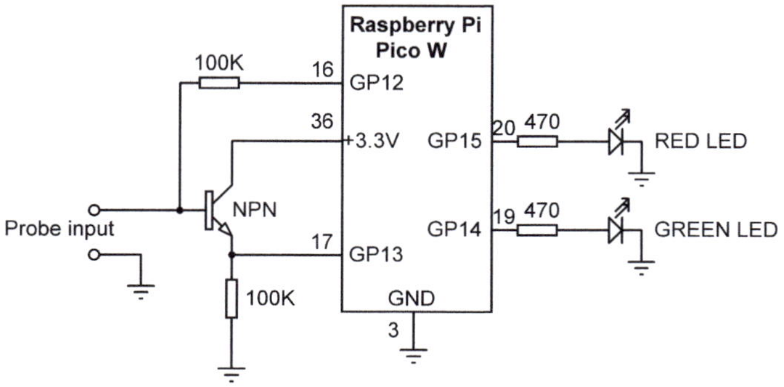

Figure 3.50: Modified circuit diagram.

The transistor is configured as an emitter-follower stage with the base controlled from output pin GP12 of the microcontroller through a 100K resistor. The external unknown digital signal is applied to the base of the transistor. The emitter of the transistor is connected to input pin GP13 of the microcontroller through a 100K resistor. Pin GP12 applies logic levels to the base of the transistor and then pin GP13 determines the state of the external signal (probe) as shown in Table 3.4. For example, if after setting GP12 = 1, you detect GP13 =1 and also after setting GP12 = 0 you again detect GP13 = 1 then the probe must be at logic 1.

Probe State	Applied to GP12	Detected at GP13	GREEN LED	RED LED
Probe at High-impedance	1	1	OFF	OFF
	0	0		
Probe at logic 1	1	1	ON	OFF
	0	1		
Probe at logic 0	1	0	OFF	ON
	0	0		

Table 3.4: Applied and detected logic levels.

Program listing: Figure 3.51 shows the program listing (Program: **probe2**). The program sends signals from TSTOUT (GP12) and checks the inputs as shown in Table 3.4. The RED and GREEN LEDs are controlled as described in the Table.

```
#--------------------------------------------------------
#              ADVANCED LOGIC PROBE
#              ====================
#
# This is a logic probe program. The state (0 or 1) of an
# unknown digital signal is indicated on two LEDs. In this
# version of the program high-impedance state is also detected
#
# Author: Dogan Ibrahim
```

```
# File   : probe2.py
# Date   : October, 2022
#-----------------------------------------------------------
from machine import Pin

RED = Pin(15, Pin.OUT)                          # RED LED
GREEN = Pin(14, Pin.OUT)                        # GREEN LED
TSTIN = Pin(13, Pin.IN)                         # EMITTER
TSTOUT = Pin(12, Pin.OUT)                       # BASE

while True:                                     # DO FOREVER
    TSTOUT.value(1)
    if TSTIN.value() ==1:
        TSTOUT.value(0)
        if TSTIN.value() == 0:
            GREEN.value(0)
            RED.value(0)
        else:
            RED.value(0)
            GREEN.value(1)
    else:
         TSTOUT.value(0)
         if TSTIN.value() == 0:
             GREEN.value(0)
             RED.value(1)
```

Figure 3.51: Program: ***probe2****.*

Chapter 4 • Raspberry Pi Pico W Multi-Digit 7-Segment Display Projects

4.1 Overview

Multi-digit 7-segment displays require continuous refreshing of their digits so that the human eye sees the digits as steady and non-flashing. The general technique used is to enable each digit for a short time (e.g. 10 ms) so that the human eye sees both digits ON at any time. This process requires the digits to be enabled alternately and continuously. As a result of this, the processor cannot perform any other tasks and has to be busy all the time, refreshing the digits. One technique used in non-multitasking systems is to use timer interrupts and refresh the digits in the timer interrupt service routines. In this chapter, you will be employing a multitasking approach to refresh the display digits so that the processor can carry out other tasks.

4.2 7-Segment LED Displays

Displaying data is one of the fundamental output activities of any microcontroller system. For example, displays are used to show the sensor data such as the temperature, humidity, pressure etc. There are several types of display devices that can be used in microcontroller-based systems. LCDs and 7-segment displays are probably two of the most commonly used display devices. There are several types of LCDs, such as text-based LCDs, graphics LCDs, color LCDs, touch screen LCDs etc. 7-segment displays are used to display numeric or alphanumeric values, and they can have one or more digits. One-digit displays can only display numbers from 0 to 9. Two-digit displays can display numbers from 0 to 99, three-digit displays numbers from 0 to 999, and so on.

As shown in Figure 4.1, a 7-segment LED display basically consists of 7 LEDs connected such that numbers from 0 to 9 and some letters can be displayed. Segments are identified by letters from **a** to **g**. Figure 4.2 shows the segment names of a typical 7-segment display.

Figure 4.1: Some 7-segment displays.

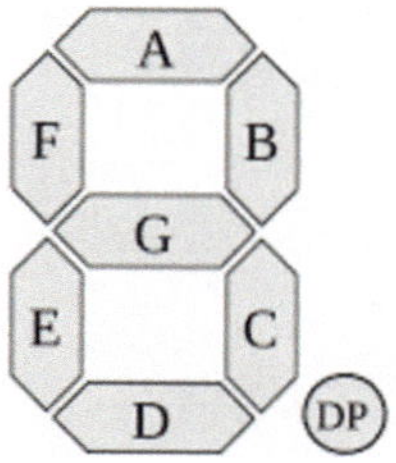

Figure 4.2: Segment names of a 7-segment display.

Figure 4.3 shows how numbers from 0 to 9 can be obtained by turning ON or OFF different segments of the display.

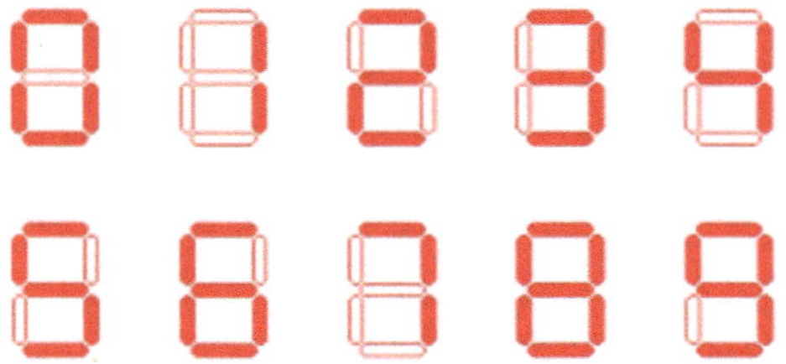

Figure 4.3: Displaying numbers 0 – 9.

7-segment LED displays are available in two different configurations: **common-cathode** and **common-anode**. As shown in Figure 4.4, in common-cathode configuration all the cathodes of all the segment LEDs are connected together to ground. The segments are then turned ON by applying a logic 1 to the required segment LED via current-limiting resistors. In common-cathode configuration, the 7-segment LED is connected to the microcontroller in current-sourcing mode.

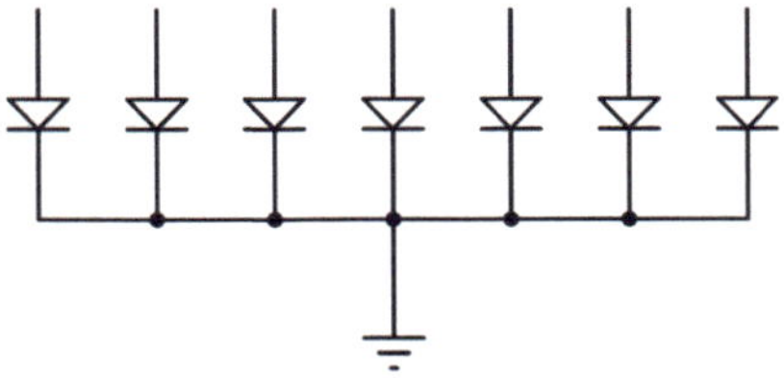

Figure 4.4: Common-cathode 7-segment LED display.

In a common-anode configuration, the anode terminals of all the LEDs are connected together as shown in Figure 4.5. This common point is then normally connected to the supply voltage. A segment is turned ON by connecting its cathode terminal to logic 0 via a current-limiting resistor. In common-anode configuration, the 7-segment LED is connected to the microcontroller in current-sinking mode.

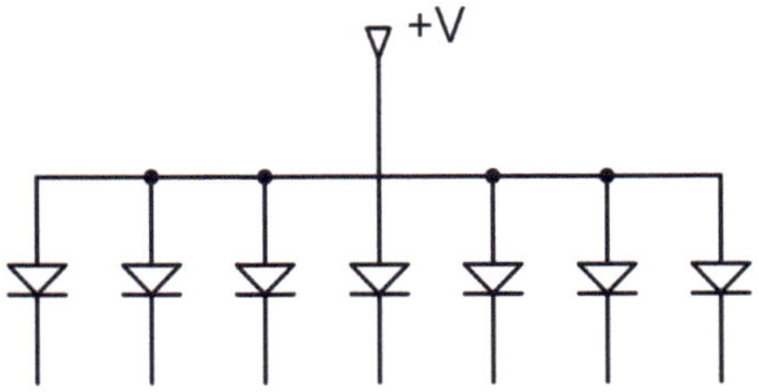

Figure 4.5: Common-anode 7-segment LED display.

In multiplexed LED applications (for example, see Figure 4.6 for a 2-digit multiplexed LED display), the LED segments of all the digits are tied together and the common pins of each digit are turned ON separately by the microcontroller. By displaying each digit for several milliseconds the eye can not differentiate that the digits are not ON all the time. This way you can multiplex any number of 7-segment displays together. For example, to display number 57, you have to send 5 to the first digit and enable its common pin. After a few milliseconds, number 7 is sent to the second digit and the common point of the second digit is enabled. When this process is repeated continuously the user sees as if both displays are ON continuously.

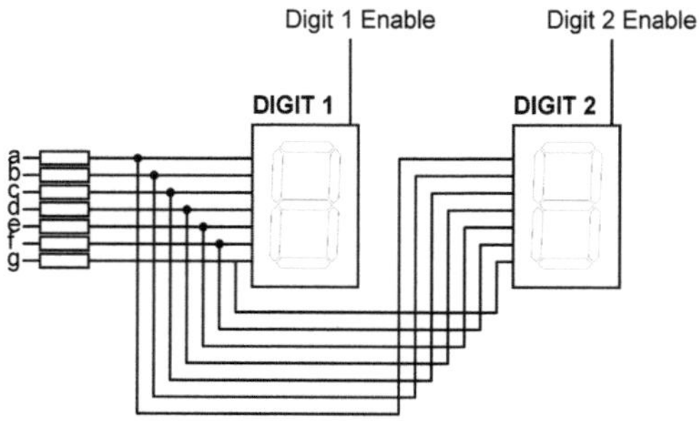

Figure 4.6: 2-digit multiplexed 7-segment LED display.

Some manufacturers provide multiplexed multi-digit displays in single packages. For example, you can purchase 2-,4-, or 8-digit multiplexed displays in a single package. The display used in this chapter is the DC56-11EWA which is a red 0.56-inch height **common-cathode** two-digit multiplexed display having 18 pins, where the pin configuration is shown in Table 4.1. Two such displays are used to make a 4-digit display. Basically, this display can be controlled from the microcontroller as follows:

- Send the segment bit pattern for digit 1 to segments a to g.
- Enable digit 1.
- Wait for a few milliseconds.
- Disable digit 1.
- Send the segment bit pattern for digit 2 to segments a to g.
- Enable digit 2.
- Wait for a few milliseconds.
- Disable digit 2.
- Wait for a few milliseconds.
- Send the segment bit pattern for digit 3 to segments a to g.
- Enable digit 3.
- Wait for a few milliseconds.
- Disable digit 3.
- Send the segment bit pattern for digit 4 to segments a to g.
- Enable digit 4.
- Wait for a few milliseconds.
- Disable digit 4.
- Repeat the above process continuously.

Pin no(s)	Segment
1,5	e
2,6	d
3,8	c
14	digit 1 Enable
17,7	g
15,10	b
16,11	a
18,12	f
13	digit 2 Enable
4	decimal Point1
9	decimal Point 2

Table 4.1: Pin configuration of DC56-11EWA dual display.

The segment configuration of the DC56-11EWA display is shown in Figure 4.7. In a multiplexed display application, the segment pins of corresponding segments are connected together. For example, pins 11 and 16 are connected as the common **a** segment. Similarly, pins 15 and 10 are connected as the common **b** segment, and so on.

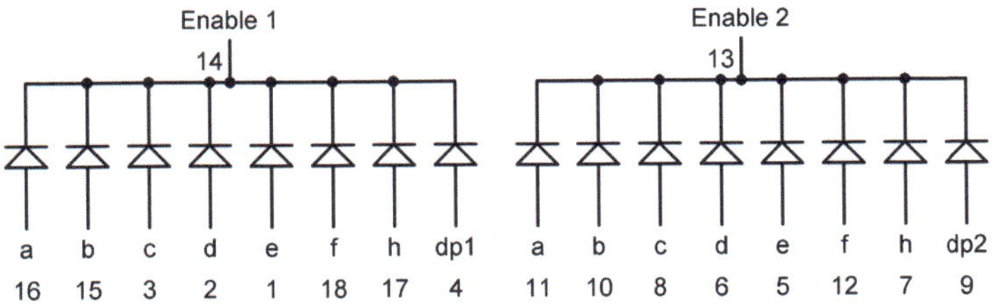

Figure 4.7: DC56-11EWA display segment configuration.

4.3 Project 1: 4-digit 7-segment display seconds counter

Description: In this project, a 7-segment 4-digit multiplexed LED display is used as a counter to count up every second from 0 to 9999.

The operation of a 4-digit multiplexed display is such that the LED segments of all the digits are tied together and the common pins of each digit are turned ON separately by the microcontroller. By displaying each digit for several milliseconds, the eye can not differentiate that the digits are not ON all the time. This way, you can multiplex any number of 7-segment displays together. For example, to display number 5734, you have to send 5 to the first digit and enable its common pin. After a few milliseconds, number 7 is sent to the second digit and the common point of the second digit is enabled and so on. When this process is repeated continuously, the user sees as if both displays are ON continuously.

The display used in this project is the DC56-11EWA which is a red 0.56-inch height common-cathode two-digit multiplexed display having 18 pins, where the pin configuration is shown in Table 4.1. Two such display modules are used to construct a 4-digit display. Each module has its own E1 and E2 enable pins.

In a multiplexed display application, the segment pins of corresponding segments are connected together. For example, pins 11 and 16 are connected as the common **a** segment. Similarly, pins 15 and 10 are connected as the common **b** segment and so on.

Block Diagram: Figure 4.8 shows the block diagram of the project.

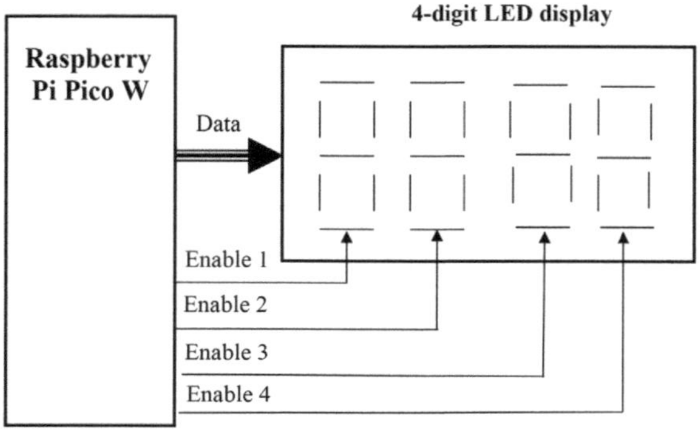

Figure 4.8: Block diagram of the project.

Circuit Diagram: The circuit diagram of the project is shown in Figure 4.9. In this project, the following pins of the Raspberry Pi Pico W are used to interface with the 7-segment LED display:

7-Segment Display pin	Raspberry Pi Pico W GPIO	Physical pin no
a	0	1
b	1	2
c	2	4
d	3	5
e	4	6
f	5	7
g	6	9
E1	7 (via transistor)	10
E2	8 (via transistor)	11
E1	9 (via transistor)	12
E2	10 (via transistor)	14

7-segment display segments are driven from the port pins through 470 Ohm current-limiting resistors. Digit enable pins E1, E2 of the first module and E1, E2 of the second module are driven from port pins GP7, GP8, GP9, and GP10 respectively through four BC108 type NPN transistors (any other NPN transistor can be used here), used as switches. The collectors of these transistors drive the segment digits. The segments are enabled when the base of the corresponding transistor is set to logic 1. Notice that the following pins of the display are connected together to form a multiplexed display:

16 and 11, 15 and 10, 3 and 8, 2 and 6, 1 and 5, 17 and 7, 18 and 12

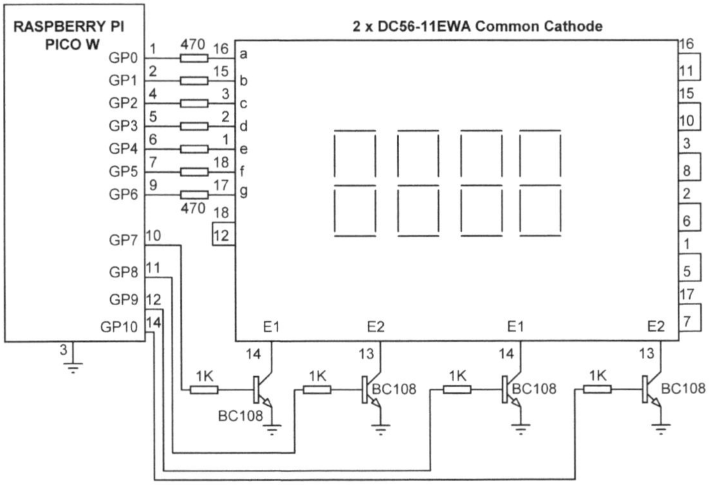

Figure 4.9: Circuit diagram of the project.

Program Listing: Before driving the display, you have to know the relationship between the numbers to be displayed and the corresponding segments to be turned ON, and this is shown below:

Number to be displayed	**LED bit pattern (a,b,c,d,e,f,g)**
1.	1,1,1,1,1,1,0
2.	0,1,1,0,0,0,0
3.	1,1,0,1,1,0,1
4.	1,1,1,1,0,0,1
5.	0,1,1,0,0,1,1
6.	1,0,1,1,0,1,1
7.	1,0,1,1,1,1,1
8.	1,1,1,0,0,0,0
9.	1,1,1,1,1,1,1
10.	1,1,1,1,0,1,1

Figure 4.10 shows the program listing (Program: **SevenCount4**). At the beginning of the program, the modules used in the program are imported. Here, the list **LED_Digits** contains 4 numbers which are the digit enable pins of the four 7-segment LED modules. Function **Refresh** is called by the timer interrupt to refresh the display. If the number is less than 4 digits, then spaces are inserted in front of it to blank the display for these digit positions. Notice that the digit count runs from 0 to 4 and not from 0 to 2 which was the case with the 2-digit display. The delay between the digit enables is reduced to 5 ms since you have 4 digits and a faster refresh rate is required. Variable **count** starts from 0 at the

beginning of the program by default. It is incremented by 1 every second. When the count reaches 10000, the timer interrupt is disabled and counting stops.

```
#-----------------------------------------------------------
#                               4-DIGIT 7-SEGMENT SECONDS COUNTER
#                               =================================
#
# In this program a 4-digit 7-segment display is connected
# to the Pico. The program counts up every second.
# In this version of the program leading zeros are omitted
#
# Author: Dogan Ibrahim
# File  : SevenCount4.py
# Date  : October, 2022
#------------------------------------------------------------
from machine import Pin, Timer
import utime

tim = Timer()
LED_Segments = [0, 1 ,2, 3, 4, 5, 6]
LED_Digits = [7, 8, 9, 10]
L = [0]*7
D = [0, 0, 0, 0]

#
# LED bit pattern for all numbers 0-9
#
LED_Bits ={
' ':(0,0,0,0,0,0,0),            # Blank
'0':(1,1,1,1,1,1,0),            # 0
'1':(0,1,1,0,0,0,0),            # 1
'2':(1,1,0,1,1,0,1),            # 2
'3':(1,1,1,1,0,0,1),            # 3
'4':(0,1,1,0,0,1,1),            # 4
'5':(1,0,1,1,0,1,1),            # 5
'6':(1,0,1,1,1,1,1),            # 6
'7':(1,1,1,0,0,0,0),            # 7
'8':(1,1,1,1,1,1,1),            # 8
'9':(1,1,1,1,0,1,1)}            # 9

count = 0                       # Initialzie count

#
# This function configures the LED ports as outputs
#
def Configure_Port():
```

```
   for i in range(0, 7):
         L[i] = Pin(LED_Segments[i], Pin.OUT)

   for i in range(0, 4):
         D[i] = Pin(LED_Digits[i], Pin.OUT)


#
# Refresh the 7-segment display
#
def Refresh(timer):                              # Thread Refresh
   global count
   cnt = str(count)                              # into string
   if len(cnt) == 3:                             # 3 digits?
      cnt = " " + cnt                            # Make sure 4 digits
   elif len(cnt) == 2:                           # 2 digits?
       cnt = "  " + cnt                          # MAke sure 4 digits
   elif len(cnt) == 1:                           # 1 digit?
       cnt = "   " + cnt                         # MAke sure 4 digits
   for dig in range(4):                          # Do for 4 digits
      for loop in range(0,7):
         L[loop].value(LED_Bits[cnt[dig]][loop])
      D[dig].value(1)
      utime.sleep(0.005)
      D[dig].value(0)

#
# Configure PORT to all outputs
#
Configure_Port()

#
# Main program loop. Start the periodic timer and counting
#
tim.init(freq=50, mode=Timer.PERIODIC, callback=Refresh)

while True:                              # Do forever
     utime.sleep(1)                      # Wait a second
     count = count + 1                   # Increment count
     if count == 10000:                  # If count = 10000 stop
       tim.deinit()                      # Stop timer
```

Figure 4.10: Program: **SevenCount4***.*

4.4 Project 2: 4-digit 7-segment display conveyor belt goods counter

Description: In this project, you're counting the number of items (e.g. bottles) passing on a conveyor belt and display the result continuously on the 7-segment display.

Block diagram: Figure 4.11 shows the system. It is assumed that you wish to count the number of bottles passing on a conveyor belt. A light beam is directed at a point to the passing bottles. On the other side of the beam, a light-dependent resistor (LDR) is used to detect when the light beam is interrupted. When this happens, a signal is sent to the development board which then increments a counter and displays the total count on the display.

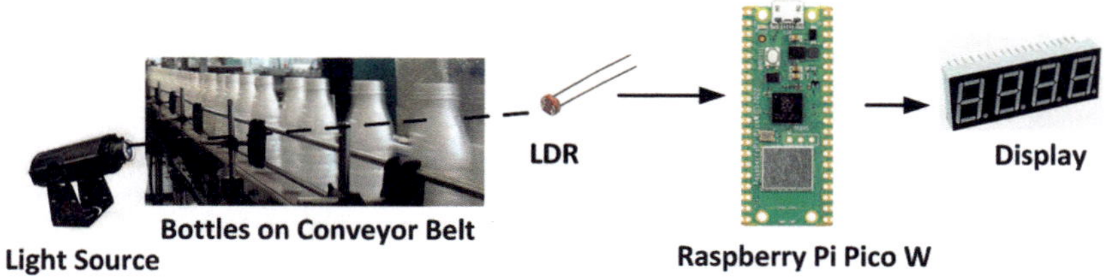

Figure 4.11: Block diagram of the project.

Light-dependent resistor

A light-dependent resistor is simply a resistor whose resistance changes with the application of light to its surface (Figure 4.12). The resistance of an LDR increases as the light intensity falling on the device is reduced.

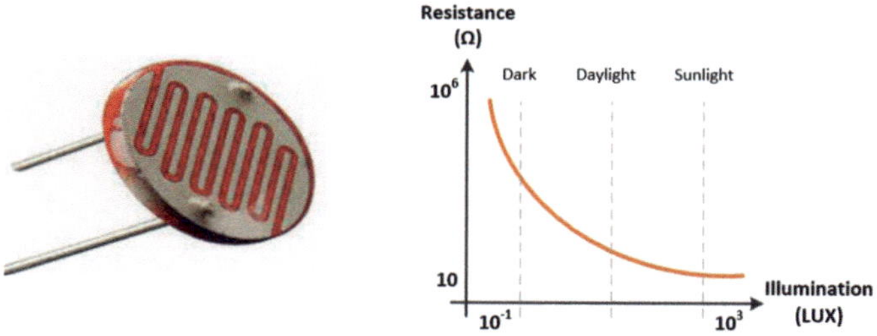

Figure 4.12: LDR and its typical characteristic.

LDRs are usually used in series with resistors in a circuit to form a resistive potential divider circuit. The voltage at the output of the potential divider circuit is used to send a trigger signal when the light intensity is below (or above) a set level. The trigger point can be detected by an MCU using analog or digital inputs.

Circuit diagram: Figure 4.13 shows the circuit diagram of the project. A 5-k-ohms potentiometer is used in series with the LDR. The potentiometer is adjusted such that the output voltage goes over 3 V when the light falling on the LDR is reduced by a passing bottle on the conveyor belt. The output of the potential divider circuit is fed to the GP11 port pin of the Raspberry Pi Pico W development board.

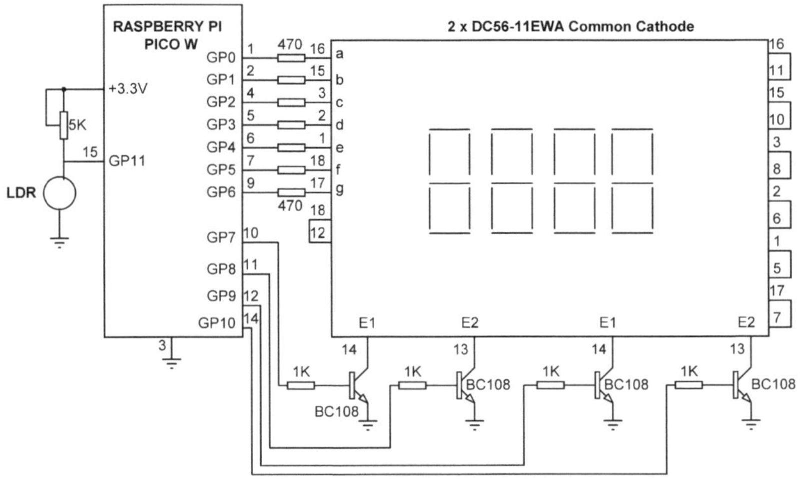

Figure 4.13: Circuit diagram of the project.

It was measured by the author that the resistance of the supplied LDR is about 0.5 k-ohms in light and increases to about 100 k-ohms when in dark. Assuming the potentiometer is set to its middle arm point, i.e., 2.5 k-ohms:

The voltage at the output when light falls on the LDR is

Vo = 3.3 V × 0.5 k-ohms / (2.5 k-ohms + 0.5 k-ohms) = 0.55 V

which is at logic 0.

Similarly, the voltage at the output when it is dark is:

Vo = 3.3 V × 100 k-ohms / (2.5 k-ohms + 100 k-ohms) = 3.21 V

which is at logic 1.

You will set the potentiometer arm to just below its mid-point so that the output voltage is even lower than 0.55 V when it is light. The exact point can easily be determined by experimentation.

Program listing: Figure 4.14 shows the program listing (Program: **Conveyor**). Most parts of this program are the same as the previous program. Here, LDR is assigned to port GP11. Inside the main program loop, the output of the LDR is read and the program waits if it is 0 (i.e. no bottles in front of the LDR). When there is a bottle in front of the LDR, its output goes to logic 1. This is detected by the program and the variable **count** is incremented by 1. The program waits until the bottle has passed the face of the LDR.

```
#-----------------------------------------------------------
#                        CONVEYOR BELT GOODS COUNTER
#                        ===========================
#
# In this program a 4-digit 7-segment display is connected
# to the Pico. Also an LCD is used. The program counts the
# goods passing on a conveyor belt and displays the total
#
# Author: Dogan Ibrahim
# File  : Conveyor.py
# Date  : October, 2022
#------------------------------------------------------------
from machine import Pin, Timer
import utime

tim = Timer()
LED_Segments = [0, 1 ,2, 3, 4, 5, 6]
LED_Digits = [7, 8, 9, 10]
L = [0]*7
D = [0, 0, 0, 0]
LDR = Pin(11, Pin.IN)

#
# LED bit pattern for all numbers 0-9
#
LED_Bits ={
' ':(0,0,0,0,0,0,0),            # Blank
'0':(1,1,1,1,1,1,0),            # 0
'1':(0,1,1,0,0,0,0),            # 1
'2':(1,1,0,1,1,0,1),            # 2
'3':(1,1,1,1,0,0,1),            # 3
'4':(0,1,1,0,0,1,1),            # 4
'5':(1,0,1,1,0,1,1),            # 5
'6':(1,0,1,1,1,1,1),            # 6
'7':(1,1,1,0,0,0,0),            # 7
'8':(1,1,1,1,1,1,1),            # 8
'9':(1,1,1,1,0,1,1)}            # 9

count = 0                       # Initialzie count

#
# This function configures the LED ports as outputs
#
def Configure_Port():
   for i in range(0, 7):
         L[i] = Pin(LED_Segments[i], Pin.OUT)
```

```
    for i in range(0, 4):
          D[i] = Pin(LED_Digits[i], Pin.OUT)


#
# Refresh the 7-segment display
#
def Refresh(timer):                                 # Thread Refresh
    global count
    cnt = str(count)                                # into string
    if len(cnt) == 3:                               # 3 digits?
       cnt = " " + cnt                              # Make sure 4 digits
    elif len(cnt) == 2:                             # 2 digits?
        cnt = "  " + cnt                            # MAke sure 4 digits
    elif len(cnt) == 1:                             # 1 digit?
        cnt = "   " + cnt                           # MAke sure 4 digits
    for dig in range(4):                            # Do for 4 digits
       for loop in range(0,7):
           L[loop].value(LED_Bits[cnt[dig]][loop])
       D[dig].value(1)
       utime.sleep(0.005)
       D[dig].value(0)

#
# Configure PORT to all outputs
#
Configure_Port()

#
# Main program loop. Start the periodic timer and counting
#
tim.init(freq=50, mode=Timer.PERIODIC, callback=Refresh)

while True:                                 # Do forever
     while LDR.value() == 0:                # Wait for dark
        pass
     count = count + 1                      # Increment count
     utime.sleep(0.1)
     while LDR.value() == 1:                # Wait until light again
        pass
```

Figure 4.14: Program: **Conveyor**.

Chapter 5 • Raspberry Pi Pico W LCD Projects

5.1 Overview

In microcontroller-based systems, you usually want to interact with the system, for example, to enter a parameter, to change the value of a parameter, or to display the output of a measured variable. Data is usually entered into a system using a switch, a small keypad, or a fully blown keyboard. Data is usually displayed using an indicator such as one or more LEDs, 7-segment displays, or LCD-type displays. LCDs have the advantages that they can display alphanumeric as well as graphical data. Some LCDs have 40 or more character lengths with the capability to display data in several lines. Some other LCDs can be used to display graphical images (Graphical LCDs, or simply GLCDs), such as animation. Some displays are in single or in multi-colour, while some others incorporate back lighting so that they can be viewed in dimly lit conditions.

LCDs can be connected to a microcontroller either in parallel form or through the interface. Parallel LCDs (e.g. Hitachi HD44780) are connected using more than one data line and several control lines and the data is transferred in parallel form. It is common to use either 4 or 8 data lines and two or more control lines. Using a 4-wire connection saves I/O pins but it is slower since the data is transferred in two stages. -based LCDs on the other hand are connected to a microcontroller using only 2 wires, the data and the clock. -based LCDs are in general much easier to use and require less wiring, but they cost more than the parallel ones. In this chapter, you will be learning to use both the parallel and -based LCDs in projects.

The programming of LCDs is a complex task and requires a good understanding of the internal operations of the LCD controllers, including knowledge of their exact timing requirements. Fortunately, there are several libraries that can be used to simplify the use of both the parallel and serial LCDs.

5.2 HD44780 parallel LCD module

Although there are several types of LCDs, the type HD44780 is currently one of the most popular LCD modules used in the industry and also by hobbyists. This module is an alphanumeric monochrome display and comes in different sizes. Modules with 16 columns are popular in most small applications, but other modules with 8, 20, 24, 32, or 40 columns are also available. Although most LCDs have two lines (or rows) as the standard, it is possible to purchase models with 1 or 4 lines. LCD displays are available with standard 14-pin connectors, although 16-pin modules are also available, providing terminals for backlighting. Table 3.4 gives the pin configuration and pin functions of a 16-pin LCD module. A brief summary of the pin functions is given below:

Pin no	Name	Function
1	VSS	Ground
2	VDD	+ ve supply
3	VEE	Contrast
4	RS	Register select
5	R/W	Read/write
6	E	Enable
7	D0	Daat bit 0
8	D1	Data bit 1
9	D2	Data bit 2
10	D3	Data bit 3
11	D4	Data bit 4
12	D5	Data bit 5
13	D6	Data bit 6
14	D7	Data bit 7
15	A	Backlight anode (+)
16	K	Backlight cathode (GND)

Table 5.1: Pin configuration of HD44780 LCD module.

V_{SS} (pin 1) and VDD (pin 2) are the ground and power supply pins. The power supply should be +5 V.

VEE is pin 3 and this is the contrast control pin used to adjust the contrast of the display. The arm of a 10 k-ohms potentiometer is normally connected to this pin and the other two terminals of the potentiometer are connected to the ground and power supply pins. The contrast of the display is adjusted by rotating the potentiometer arm.

Pin 4 is the Register Select (RS) and when this pin is LOW, data transferred to the display is treated as commands. When RS is HIGH, character data can be transferred to and from the display.

Pin 5 is the Read/Write (R/W) line. This pin is pulled LOW in order to write commands or character data to the LCD module. When this pin is HIGH, character data or status information can be read from the module. This pin is normally connected permanently to LOW so that commands or character data can be sent to the LCD module.

Enable (E) is pin 6 which is used to initiate the transfer of commands or data between the LCD module and the microcontroller. When writing to the display, data is transferred only on the HIGH to LOW transition of this pin. When reading from the display, data becomes available after the LOW to HIGH transition of the enable pin and this data remains valid as long as the enable pin is at logic HIGH.

Pins 7 to 14 are the eight data bus lines (D0 to D7). Data can be transferred between the microcontroller and the LCD module using either a single 8-bit byte, or as two 4-bit nibbles. In the latter case, only the upper four data lines (D4 to D7) are used. The 4-bit mode has the advantage that four fewer I/O lines are required to communicate with the LCD. 4-bit mode is however slower since the data is transferred in two stages. In this book, you shall be using the 4-bit interface only.

Pins 15 and 16 are for background brightness control. To enable the background brightness, a 220-ohms resistor should be connected from pin 15 to the +5 V supply, and pin 16 should be connected to ground.

In 4-bit mode, the following pins of the LCD are used. The R/W line is permanently connected to ground. This mode uses 6 GPIO port pins of the microcontroller:

VSS, VDD, VEE, E, R/S, D4, D5, D6, D7

In the next sections of this chapter, you will be developing projects using LCDs. But before looking at the details of the LCD, it is worthwhile to review the basic details of the bus.

5.3 The I²C Bus

The I²C bus is one of the most commonly used microcontroller communication protocols for communicating with external devices such as sensors and actuators. The bus is a single master, multiple slave bus and it can operate at standard mode: 100 Kbit/s, full speed: 400 Kbit/s, fast mode: 1 Mbit/s, and high speed: 3.2 Mbit/s. The bus consists of two open-drain wires, pulled-up with resistors:

SDA: data line
SCL: clock line

Figure 5.1 shows the structure of an I²C bus with one master and three slaves.

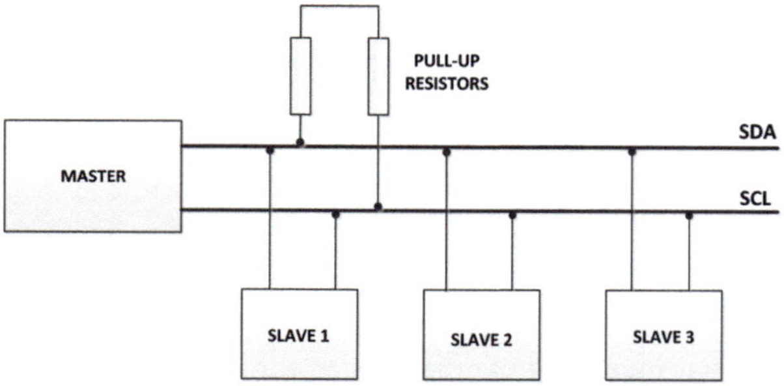

Figure 5.1: I²C bus with one master and three slaves.

Because the I²C bus is based on just two wires, there should be a way to address an individual slave device on the same bus. For this reason, the protocol defines that each slave device provides a unique slave address for the given bus. This address is usually 7-bits wide. When the bus is free, both lines are HIGH. All communication on the bus is initiated and completed by the master which initially sends a START bit, and completes a transaction by sending a STOP bit. This alerts all the slaves that some data is coming on the bus and all the slaves listen on the bus. After the start bit, 7 bits of the unique slave address are sent. Each slave device on the bus has its own address and this ensures that only the addressed slave communicates on the bus at any time to avoid any collisions. The last sent bit is a read/write bit such that if this bit is 0, it means that the master wishes to write to the bus (e.g. to a register of a slave), if this bit is a 1, it means that the master wishes to read from the bus (e.g. from the register of a slave). The data is sent on the bus with the MSB bit first. An acknowledgment (ACK) bit takes place after every byte and this bit allows the receiver to signal to the transmitter that the byte was received successfully and as a result, another byte may be sent. ACK bit is sent at the 9th clock pulse.

The communication over the I^2C bus is simply as follows:

- The master sends on the bus the address of the slave it wants to communicate with.
- The LSB is the R/W bit which establishes the direction of data transmission, i.e. from master to slave (R/W = 0), or from slave to master (R/W = 1).
- Required bytes are sent, each interleaved with an ACK bit, until a STOP condition occurs.

Depending on the type of slave device used, some transactions may require a separate transaction. For example, the steps to read data from an -compatible memory device are:

- Master starts the transaction in write mode (R/W = 0) by sending the slave address on the bus.
- The memory location to be retrieved is then sent as two bytes (assuming 64-Kbit memory).
- The master sends a STOP condition to end the transaction.
- The master starts a new transaction in read mode (R/W = 1) by sending the slave address on the bus.
- The master reads the data from the memory. If reading the memory in sequential format, then more than one byte will be read.
- The master sets a STOP condition on the bus.

5.4 pins of the Raspberry Pi Pico W

Raspberry Pi Pico and Pico W have two I^2C pins, named I2C0 and I2C1 (see Figure 5.2). As shown in the figure, the pins are duplicated and shared with other pins. For example, GP0 (pin 1) is the I2C0 SDA pin and GP1 (pin 1) is the I2C0 SCL pin. Also, GP16 (pin 21) is the I2C0 SDA pin and GP17 (pin 22) is the I2C SCL pin.

Figure 5.2: Raspberry Pi Pico and Pico W I²C pins.

The default pins are:

I2C0 SCL	GP9
I2C0 SDA	GP8
I2C1 SCL	GP7
I2C1 SDA	GP6

5.5 Project 1: Parallel LCD – displaying text

Description: In this project, you will develop a number of functions that can be used to send data and text to 16 × 2-character parallel LCD displays.

Aim: The aim of this project is to develop a library of functions that can be used to control parallel LCDs. These functions can be used in projects to send text and numbers to LCDs.

Block diagram: Figure 5.3 shows the block diagram of the project.

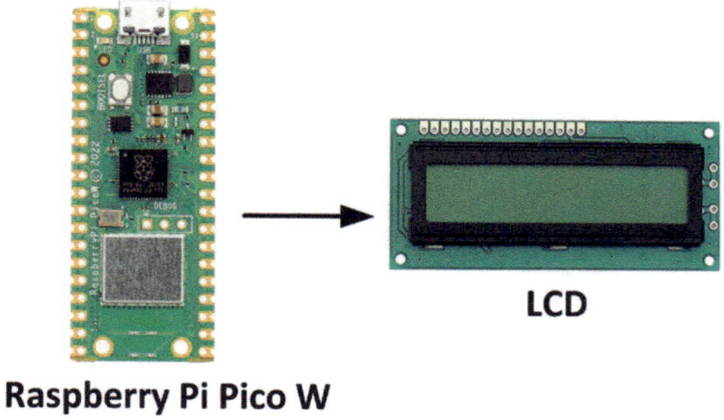

Figure 5.3: Block diagram of the project.

Circuit diagram: The circuit diagram of the project is shown in Figure 5.4. The LCD is connected to the Pico using four data wires (D4 – D7) and two control wires (E and R/S). The connections between the LCD and Pico are as follows:

LCD pin	**Pico pin**
E	GP0
R/S	GP1
D4	GP2
D5	GP3
D6	GP4
D7	GP5

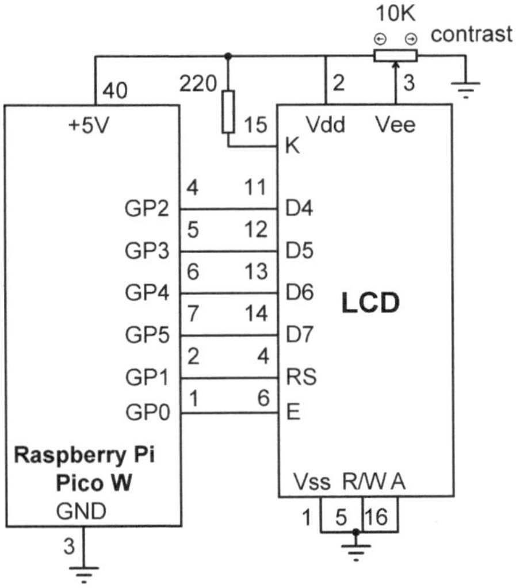

Figure 5.4: Circuit diagram of the project.

The contrast of the LCD is controlled using a 10-kOhms potentiometer. Notice that, the LCD is operated with +5 V. Pico I/O ports is not +5 V compatible, but this is not a problem for the Pico because all pins of the LCD are in input mode (**make sure that the R/W pin is connected to GND before powering up the circuit**).

Program listing: Figure 5.5 shows the functions listing (Program: **LCD**). The connections between the LCD and Pico are defined at the beginning and these can be changed if desired. The remainder of the functions should not be changed for proper control of the LCD. These functions implement the initialization and control of the LCD.

The following LCD control functions are available:

lcd_init: this is the LCD initialization function and must be called first before any other functions are called

lcd_clear: clears the LCD

lcd_home: homes the cursor (top left position)

lcd_cursor_blink: enables blinking cursor

lcd_cursor_on: enables visible cursor

lcd_cursor_off: disables visible cursor

lcd_puts(s): displays string s

lcd_putch(c): displays character c

lcd_goto(col,row): positions the cursor at the specified column and position. (0, 0) is the left corner of the LCD. The first row is row 0, the second row is row 1, and so on.

```
#----------------------------------------------------------
#                      PARALLEL LCD FUNCTIONS
#                      ======================
#
# These functions initialize and control the paralell LCD
#
# Author: Dogan Ibrahim
# File  : LCD
# Date  : October 2022
#------------------------------------------------------------
from machine import Pin
import utime

EN = Pin(0, Pin.OUT)
RS = Pin(1, Pin.OUT)
D4 = Pin(2, Pin.OUT)
D5 = Pin(3, Pin.OUT)
D6 = Pin(4, Pin.OUT)
D7 = Pin(5, Pin.OUT)
PORT = [2, 3, 4, 5]
L = [0,0,0,0]

def Configure():
    for i in range(4):
       L[i] = Pin(PORT[i], Pin.OUT)

def lcd_strobe():
    EN.value(1)
    utime.sleep_ms(1)
    EN.value(0)
    utime.sleep_ms(1)
```

```
def lcd_write(c, mode):
    if mode == 0:
       d = c
    else:
       d = ord(c)
    d = d >> 4
    for i in range(4):
        b = d & 1
        L[i].value(b)
        d = d >> 1
    RS.value(mode)
    lcd_strobe()

    if mode == 0:
       d = c
    else:
       d = ord(c)
    for i in range(4):
        b = d & 1
        L[i].value(b)
        d = d >> 1
    RS.value(mode)
    lcd_strobe()
    utime.sleep_ms(1)
    RS.value(1)

def lcd_clear():
    lcd_write(0x01, 0)
    utime.sleep_ms(5)

def lcd_home():
    lcd_write(0x02, 0)
    utime.sleep_ms(5)

def lcd_cursor_blink():
    lcd_write(0x0D, 0)
    utime.sleep_ms(1)

def lcd_cursor_on():
    lcd_write(0x0E, 0)
    utime.sleep_ms(1)

def lcd_cursor_off():
    lcd_write(0x0C, 0)
    utime.sleep_ms(1)
```

```
def lcd_puts(s):
    l = len(s)
    for i in range(l):
       lcd_putch(s[i])

def lcd_putch(c):
    lcd_write(c, 1)

def lcd_goto(col, row):
    c = col + 1
    if row == 0:
        address = 0
    if row == 1:
        address = 0x40
    address = address + c - 1
    lcd_write(0x80 | address, 0)

def lcd_init():
    Configure()
    utime.sleep_ms(120)
    for i in range(4):
        L[i].value(0)
    utime.sleep_ms(50)
    L[0].value(1)
    L[1].value(1)
    lcd_strobe()
    utime.sleep_ms(10)
    lcd_strobe()
    utime.sleep_ms(10)
    lcd_strobe()
    utime.sleep_ms(10)
    L[0].value(0)
    lcd_strobe()
    utime.sleep_ms(5)
    lcd_write(0x28, 0)
    utime.sleep_ms(1)
    lcd_write(0x08, 0)
    utime.sleep_ms(1)
    lcd_write(0x01, 0)
    utime.sleep_ms(10)
    lcd_write(0x06, 0)
    utime.sleep_ms(5)
    lcd_write(0x0C, 0)
    utime.sleep_ms(10)
#================= END OF LCD FUNCTIONS ======================
```

```
lcd_init()
lcd_puts('Hello from PICO')
```

Figure 5.5: Program: **LCD***.*

Displaying text

The following statement displays text **Hello from PICO** on the LCD:

lcd_init()

lcd_puts('Hello from PICO')

Figure 5.6 shows the display.

Figure 5.6: Displaying text.

Storing the LCD functions in a module library

You can easily combine all the LCD functions in a library and then import this library at the beginning of your program. The steps are given below:

- Open program **LCD** (see Figure 5.5).
- Delete the last two statements.
- Click at Thonny: **File** followed by **Save As**.
- Select **Raspberry Pi Pico** as the destination.
- Set the filename as **LCDFuncs.py** and click **OK**.

You can now import library **LCDFuncs** into your LCD based programs. For example, the program given in this project can be rewritten as shown in Figure 5.7 (Program: **LCD2**). Notice that all the LCD functions must be preceded with word **LCDFuncs**.

```
#----------------------------------------------------------
#                          PARALLEL LCD FUNCTIONS
#                          ======================
#
# These functions initialize and control the paralell LCD
#
# Author: Dogan Ibrahim
# File  : LCD2
```

```
# Date   : October 2022
#-------------------------------------------------------------
import utime
import LCDFuncs

LCDFuncs.lcd_init()

LCDFuncs.lcd_puts('Hello again...')
```

Figure 5.7: Program: **LCD2**.

5.6 Project 2: Reaction-timer with parallel LCD

Description: This is a reaction-timer game. The idea of the game is to measure the reaction time of the user. The game consists of an LCD, an LED, and a pushbutton. The game starts with the user keeping his/her hand on the pushbutton. The LED turns ON at random times and as soon as it is ON, the user is expected to press the pushbutton. The elapsed time between seeing the LED lit and pressing the pushbutton is measured and displayed on the LCD as the reaction time of the user.

Block diagram: Figure 5.8 shows the block diagram of the project.

Figure 5.8: Block diagram of the project.

Circuit diagram: The circuit diagram of the project is shown in Figure 5.9. The LCD is connected as in the previous project. The LED and the pushbutton are connected to GP16 and GP17 respectively. An internal pull-up resistor is used at pin GP17 so that there is no need to use an external resistor.

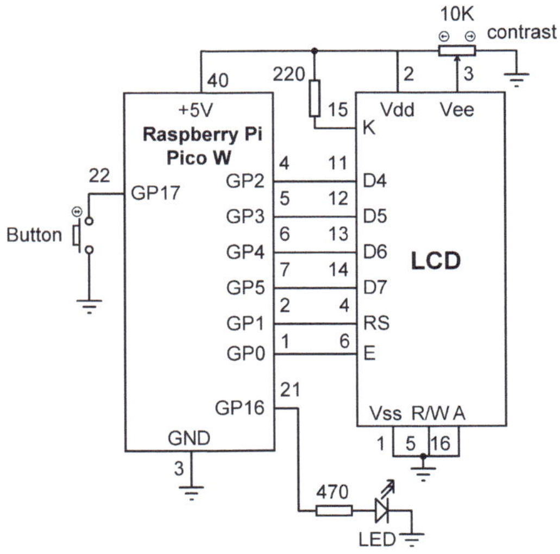

Figure 5.9: Circuit diagram of the project.

Program listing: Figure 5.10 shows the program listing (Program: **Reaction**). At the beginning of the program, the LCD module, machine, utime, and the random module are all imported to the program. **Button** and **LED** are assigned to port pins GP17 and GP16 respectively. GP17 is pulled-up internally so that the state of this pin is logic 1 by default, and goes to logic 0 when the pushbutton is pressed. LCD library is then initialized by calling function **lcd_init**.

At the beginning of the main program, the LED is turned OFF, and the program enters a **while** loop. Inside this loop, the LCD is cleared and a random number is generated between 3 and 10. This value is used to delay turning the LED ON so that the user does not know when the LED is going to be lit. At this point, the LED is turned ON, a timer is started, and the button interrupt is enabled. The interrupt is configured to be activated on the falling edge of the pushbutton output (i.e. when the pushbutton is pressed). The main program then waits until the interrupt is serviced (i.e. while the **flag** is 0).

The program jumps to function **MyButton** as soon as the pushbutton is pressed. At the beginning of this program, further interrupts are disabled by setting **handler = None**. Inside this function, the LED is turned OFF and the timer is stopped. The elapsed time is calculated by subtracting the current time from the time when the LED was turned ON. This time is converted into string and stored in variable **ReactionStr** and is displayed in the second row of the LCD in milliseconds. The text **Reaction Time:** is displayed at the first row of the LCD. As an example, if the reaction time is 500 ms, it is displayed in the following format:

Reaction Time:
500

Variable **flag** is then set to 1 so that the main program continues. At the same time, the LCD is cleared. The game re-starts as soon as the LCD is cleared.

```
#---------------------------------------------------------
#                              REACTION TIMER
#                              ==============
#
# This is a reaction timer program which measures the
# reaction of the user and displays on the LCD in ms.
# For a fast reaction time, the user should press the
# pushbutton as soon as the LED is lit
#
# Author: Dogan Ibrahim
# File  : Reaction.py
# Date  : October 2022
#-----------------------------------------------------------
import LCDFuncs
from machine import Pin
import utime
import random

Button = Pin(17, Pin.IN, Pin.PULL_UP)
LED = Pin(16, Pin.OUT)
LCDFuncs.lcd_init()
flag = 0

#
# This is the interrupt service routine. The progra jumps
# here as soon as the pushbutton is pressed
#
def MyButton(pin):
    global flag
    Button.irq(handler = None)
    LED.value(0)
    TmrEnd = utime.ticks_ms()
    ReactionTime = utime.ticks_diff(TmrEnd, TmrStart)
    ReactionStr = str(ReactionTime)
    flag = 1
    LCDFuncs.lcd_puts("Reaction Time:")
    LCDFuncs.lcd_goto(0, 1)
    LCDFuncs.lcd_puts(ReactionStr)
    utime.sleep(3)
    LCDFuncs.lcd_clear()

#
# Start of MAIN program
#
LED.value(0)
while True:
```

```
    flag = 0
    LCDFuncs.lcd_clear()
    rnd = random.randint(3, 10)
    utime.sleep(2)
    utime.sleep(rnd)
    LED.value(1)
    TmrStart = utime.ticks_ms()
    Button.irq(handler=MyButton, trigger = Pin.IRQ_FALLING)
    while flag == 0:
        pass
```

Figure 5.10: Program: **Reaction.**

5.7 Project 3: Voltmeter with parallel LCD

Description: This is a simple voltmeter project where the voltage of an external voltage source is measured and displayed on the LCD in millivolts.

Before developing the program, it is worthwhile to review the analog-to-digital (ADC) of the Raspberry Pi Pico W.

The ADC inside Pico

Most sensors in real life are analog and give analog output voltages or currents which are proportional to the measured variable. Such sensors cannot be directly connected to digital computers without using ADCs. In this chapter, you will learn how to use the ADC channels of the Raspberry Pi Pico.

Most ADCs for general-purpose applications are 8-bits or 10-bits wide, although some higher-grade professional ones are 16- or even 32-bits wide. The conversion time of an ADC is one of its important specifications. This is the time taken for the ADC to convert an analog input into digital one. The smaller the conversion time the better. Some cheaper ADCs give the converted digital data in serial format, while some more expensive professional ones provide parallel digital outputs.

The Raspberry Pi Pico W has 5 ADC channels, where 4 of them are at pins GP26, GP27, GP28, and GP29, known as analog channels 0, 1, 2, and 3. The first three channels are available at the GPIO pins, and the fourth one can be used to measure the VSYS voltage of the board. There is also a built-in ADC channel 4 which is connected internally to a temperature sensor.

The Pico's ADC has a resolution of 12-bits, thus converting an analog input voltage into 4096 (0 to 4095) levels. MicroPython, however, transforms the output into a 16-bits number, ranging from 0 to 65535.

The reference voltage of the ADC used on the Pico is +3.3 V. Using such an ADC, the resolution is 3300 mV / 65535 = 0.050 mV per bit, or 50 µV/bit. Therefore, an analog input voltage of 0.050 mV gives digital output of 00000000 0000001, 0.1 mV gives 00000000 00000010, and so on.

Aim: The aim of this project is to show how the Pico ADC channels can be used to read analog input voltage.

Circuit diagram: Figure 5.11 shows the circuit diagram of the project. In this project, the voltage to be measured in applied to analog input GP26 (pin 31, channel 0). You must make sure that the input voltage does not exceed +3.3 V. If it is required to measure higher voltages, then you can use resistive potential divider circuits at the input of the ADC.

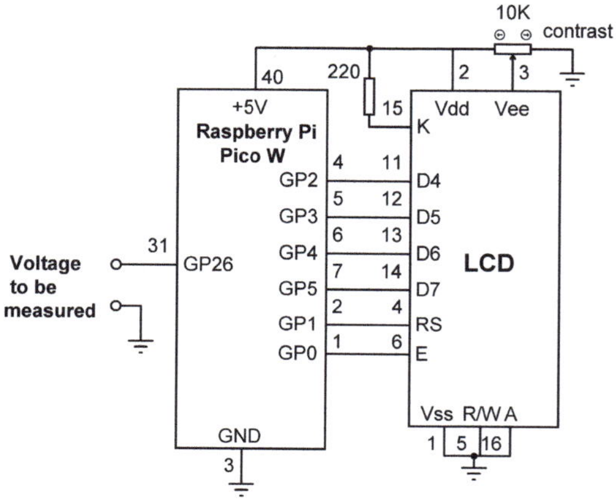

Figure 5.11: Circuit diagram of the project.

Program listing: Figure 5.12 shows the program listing and sample output from the program (Program: **VoltLCD**). At the beginning of the program module, the **ADC** of **machine** is imported to the program and variable **AnalogIn** is assigned to analog input channel 0. The conversion factor is then defined as 3300 / 65535. The value read from the analog channel must be multiplied by this number in order to calculate the actual value of the measured voltage. The remainder of the program runs in a loop where the input voltage is read, converted to millivolts, and then displayed on the LCD.

```
#----------------------------------------------------------
#               VOLTMETER
#               =========
#
# This is a voltmeter project. The voltage to me measured
# is applied to GP26 (pin 31) of the Raspberry Pi Pico W.
# The program displays the measured voltage on the LCD
#
# Author: Dogan Ibrahim
# File  : VoltLCD.py
```

```
# Date  : October 2022
#------------------------------------------------------------
from machine import ADC
import utime
import LCDFuncs

AnalogIn = ADC(0)                        # ADC channel 0
Conv = 3300 / 65535                      # Conversion factor
LCDFuncs.lcd_init()

while True:                              # Do forever
    mV = AnalogIn.read_u16()             # Read input
    mV = mV * Conv                       # Input in mV
    LCDFuncs.lcd_clear()                 # Clear screen
    mVstr = str(mV)                      # Convert to string
    LCDFuncs.lcd_puts(mVstr)             # Display
    utime.sleep(1)                       # Wait 1 second
```

Figure 5.12: Program: **VoltLCD**.

5.8 Project 4: Temperature measurement – using the internal temperature sensor – with parallel LCD

Description: In this project, the internal temperature sensor of Pico is used, and the measured temperature is displayed on the LCD.

Aim: The aim of this project is to show how the internal temperature sensor of the Pico can be used to measure the temperature.

Program listing: The internal temperature sensor is connected to ADC channel 4. The data read is converted into degrees Celsius using the following formula:

Temperature (in degrees Celsius) = 27 - (reading - 0.706)/0.001721

The program listing is shown in Figure 5.13 (Program: **TempInt**). The program reads the data from ADC channel 4, converts it into volts, and then applies the above formula to convert it into degrees Celsius. The calculated value is then displayed on the LCD. The program repeats every second. You can check that the temperature will increase if you touch the processor with the tip of your finger.

```
#---------------------------------------------------------
#              TEMPERATURE MEASUREMENT
#              =======================
#
# This program measures the temperature using the internal
# temperature sensor of the Pico and displyas on LCD
#
```

```
# Author: Dogan Ibrahim
# File  : TempInt.py
# Date  : October 2022
#------------------------------------------------------------
from machine import ADC
import utime
import LCDFuncs

AnalogIn = ADC(4)                               # ADC channel 4
Conv = 3.3 / 65535                              # Conversion factor
LCDFuncs.lcd_init()

while True:                                     # Do forever
    V = AnalogIn.read_u16()                     # Read temp
    V = V * Conv                                # Convert to Volts
    Temp = 27 - (V - 0.706) / 0.001721          # Convert to temp
    LCDFuncs.lcd_clear()                        # Clear screen
    Tempstr = str(Temp)[:5]                     # Convert to string
    LCDFuncs.lcd_puts(Tempstr)                  # Display
    utime.sleep(1)                              # Wait 1 second
```

Figure 5.13: Program: **TempInt**.

5.9 Project 5: Temperature measurement – using an external temperature sensor and with parallel LCD

Description: In this project, an external analog temperature sensor is used to measure and display the ambient temperature on the LCD.

Aim: The aim of this project is to show how an external analog temperature can be used to measure the external temperature.

Block Diagram: Figure 5.14 shows the block diagram of the project.

Figure 5.14: Block diagram of the project.

Circuit diagram: The circuit diagram of the project is shown in Figure 5.15. In this project, a TMP36-type temperature sensor chip is used (Figure 5.16) and is connected to ADC channel 0. This chip provides an analog output voltage proportional to the measured temperature. The relationship between the measured temperature and the output voltage is given by:

T = (Vo – 500) / 10

Where **T** is the measured temperature in degrees Celsius, and **Vo** is the sensor output voltage in millivolts.

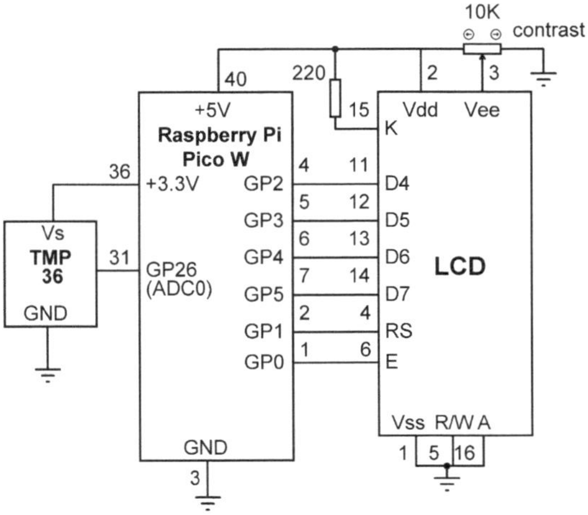

Figure 5.15: Circuit diagram of the project.

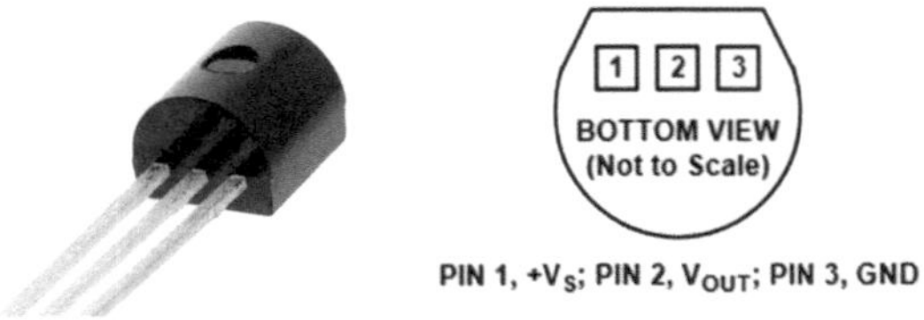

Figure 5.16: TMP36 sensor chip.

Program listing: Figure 5.17 shows the program listing (Program: **TMP36**). Sensor voltage is read by channel 0 of the ADC. This voltage is then converted into millivolts. The temperature is calculated in degrees Celsius and is displayed on the LCD every second. Notice that, the output data is formatted so that only the first five digits of string **Temp** are displayed. As an example, the temperature is displayed in the following format:

nn.mm

```
#------------------------------------------------------------
#                     TEMPERATURE MEASUREMENT
#                     =======================
#
# This program measures the temperature using an external
# TMP36 type temperature sensor chip
#
# Author: Dogan Ibrahim
# File  : TMP36.py
# Date  : October 2022
#--------------------------------------------------------------
from machine import ADC
import utime
import LCDFuncs

AnalogIn = ADC(0)                       # ADC channel 0
Conv = 3300 / 65535                     # Conversion factor
LCDFuncs.lcd_init()

while True:                             # Do forever
    V = AnalogIn.read_u16()             # Read temp
    mV = V * Conv                       # Convert to Volts
    Temp = (mV - 500.0) / 10.0          # Convert to temp
    LCDFuncs.lcd_clear()                # Clear screen
    Tempstr = str(Temp)[:5]             # Convert to string
    LCDFuncs.lcd_puts(Tempstr)          # Display
    utime.sleep(1)                      # Wait 1 second
```

Figure 5.17: Program: **TMP36**.

5.10 Project 6: ON/OFF temperature controller with parallel LCD

Description: Temperature control is important in many industrial, commercial, and in domestic chemical applications as well as in many other applications. A temperature control system basically consists of a temperature sensor, a heater, a fan (optional), an actuator to operate the heater, and a controller. A negative feedback is used to control the heater so that the temperature is at the desired set-point value. Accurate temperature control systems are based on the PID (Proportional+Integral+Derivative) algorithm which requires some knowledge about the behaviour and transfer function of the system to be controlled. In this project, an ON/OFF type simple control system is designed. ON/OFF temperature control systems commonly use relays to turn the heater ON or OFF depending on the set-point temperature and the measured temperature. If the measured temperature is below the set-point value, then the relay is activated which turns the heater ON. If on the other hand, the measured temperature is above the set-point value, then the relay is deactivated to turn OFF the heater so that the temperature is lowered.

In this project, a TMP36-type sensor chip is used together with a heater, and an LED to control the temperature of a small room. The heater is turned **ON** by the relay if the measured room temperature (**RoomTemp**) is below the set-point temperature (**SetTemp**), and it is turned **OFF** if it is above the set-point value. The LED turns ON if the room temperature is below the set point value to indicate that the heater is ON. This process is repeated every 3 seconds.

The aim: The aim of this project is to show how an ON/OFF temperature controller system can be designed using a low-cost temperature sensor chip with the Raspberry Pi Pico W.

Block diagram: Figure 5.18 shows the block diagram of the project.

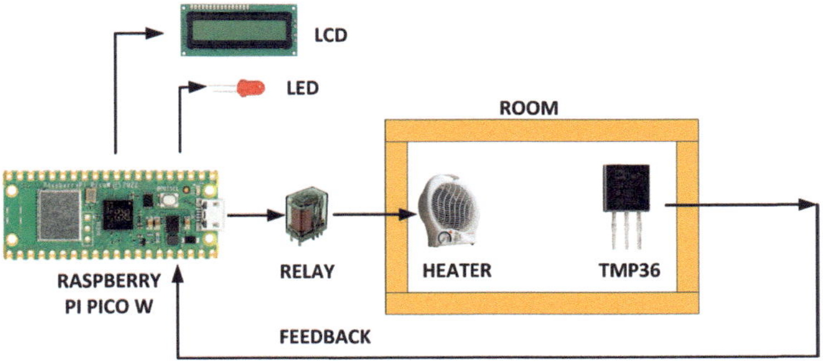

Figure 5.18: Block diagram of the project.

Circuit diagram: The circuit diagram of the project is shown in Figure 5.19. The TMP36 sensor chip is connected to analog channel 0 as in the previous project. The LED is connected to GP16 through a 470-ohms current limiting resistor. The Relay is connected to GP17 and is activated when a logic 1 (+3.3 V) is applied to it. The LCD is connected as in the previous project. The connections between the Raspberry Pi Pico W ports and various components are as follows:

Raspberry Pi Pico W	Component
GP26 (ADC0)	TMP36 out
GP16	LED
GP17	Relay
GP0-GP5	LCD

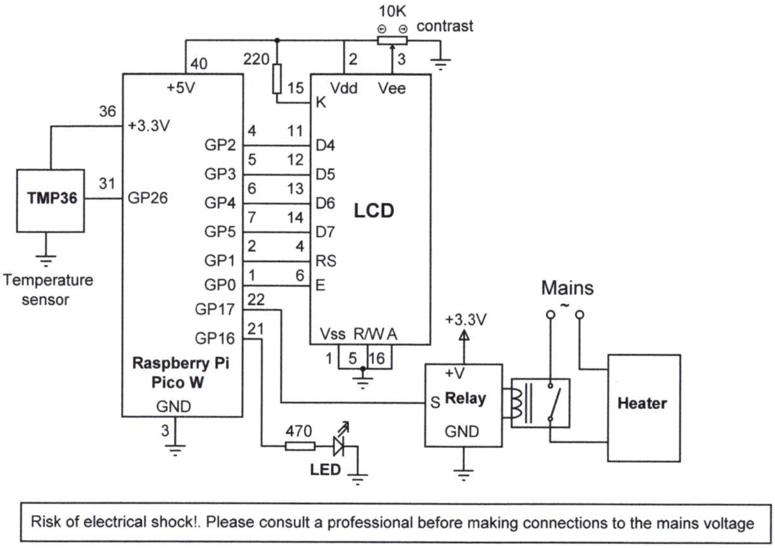

Figure 5.19: Circuit diagram of the project.

Operation of the project

The LCD shows both the desired temperature (**SetTemp**) and the measured room temperature (**RoomTemp**). The desired temperature and the room temperature are displayed in the following format, where the room temperature is updated continuously every 3 seconds:

Row 0: **Set : 23.45**
Row 1: **Meas: 22.50**

Program listing: Figure 5.20 shows the program listing (Program: **ONOFFLCD**). At the beginning of the program, the modules used are imported to the program, **SetTemp** is set to 24 °C, and the **Relay** and **LED** are configured as outputs. Inside the main program loop, the room temperature **RoomTemp** is read and compared with the **SetTemp**. If **SetTemp** is greater than **RoomTemp**, then both the **Relay** and LED are turned ON, otherwise they both are turned OFF. This loop is repeated every 3 seconds.

```
#---------------------------------------------------------
#              ON-OFF TEMPERATURE CONTROLLER
#              =============================
#
# This is an ON-OFF temperature controller program. The
# project consists of a temperature sensor, an LED and a
# heater. The heater and the LED are turned ON if the room
# temperature (RoomTemp) is below the desired value (SetTemp)
#
# An LCD is used to display the desired temperature at the top
# row, and room temperature at bottom row
```

```
#
# Author: Dogan Ibrahim
# File   : ONOFFLCD.py
# Date   : October 2022
#--------------------------------------------------------------
from machine import ADC, Pin
import utime
import LCDFuncs

LCDFuncs.lcd_init()                 # Initialize LCD
AnalogIn = ADC(0)                   # ADC channel 0
Conv = 3300 / 65535                 # Conversion factor

SetTemp = 24.0                      # Desired temperature
LED = Pin(16, Pin.OUT)              # LED at GP16
Relay = Pin(17, Pin.OUT)            # Relay at GP17
LED.value(0)                        # Turn OFF LED
Relay.value(0)                      # Turn OFF Relay

LCDFuncs.lcd_clear()                # Clear LCD
LCDFuncs.lcd_puts("Set : ")         # Display Set :
LCDFuncs.lcd_puts(str(SetTemp)[:5]) # Display SetTemp

while True:                         # Do forever
    V = AnalogIn.read_u16()         # Read temp
    mV = V * Conv                   # Convert to Volts
    RoomTemp = (mV - 500.0) / 10.0      # Measured temperature
    LCDFuncs.lcd_goto(0, 1)         # Cursor at 0,1
    LCDFuncs.lcd_puts("Meas: ")     # Display Meas:
    LCDFuncs.lcd_puts(str(RoomTemp)[:5])    # Display RoomTemp
    if RoomTemp < SetTemp:              # If Room temp < desired
        Relay.value(1)                  # Turn Relay ON
        LED.value(1)                    # Turn LED ON
    else:
        Relay.value(0)                  # Turn Relay OFF
        LED.value(0)                    # Tuen LED OFF
    utime.sleep(3)                      # Wait 3 seconds
```

Figure 5.20: Program: **ONOFFLCD**.

5.11 Project 7: Measuring the ambient light intensity – using parallel LCD

Description: In this project, a light-dependent-resistor (LDR) is used to measure and display the ambient light intensity on a parallel LCD.

Block diagram: Figure 5.21 shows the block diagram of the project.

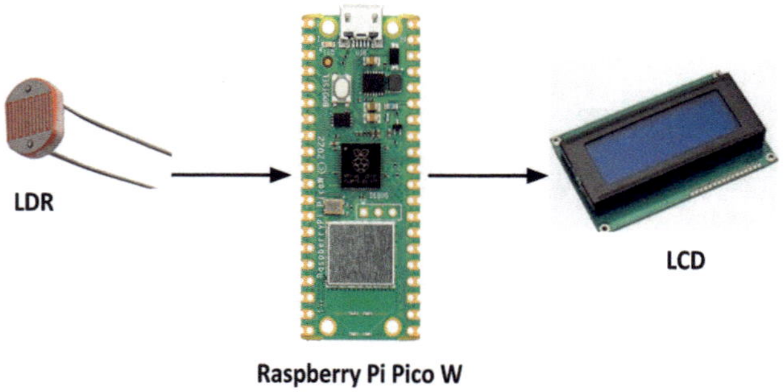

Figure 5.21: Block diagram of the project.

Background information: An LDR is an electronic component whose resistance changes with the light intensity that falls upon it. The resistance of the LDR falls with an increase in light intensity falling upon the device. Typically, the resistance in daylight could be around kilo-ohms and in dark it could be around a few megohms. As a result, you can use such a device to measure the light intensity. A typical characteristic of an LDR is shown in Figure 5.22. LDRs are used in circuits in the form of resistive potential dividers. A fixed resistor is connected in series with the LDR and the voltage across this resistor is measured. This voltage is proportional to the resistance of the LDR and hence to the light intensity falling on the LDR.

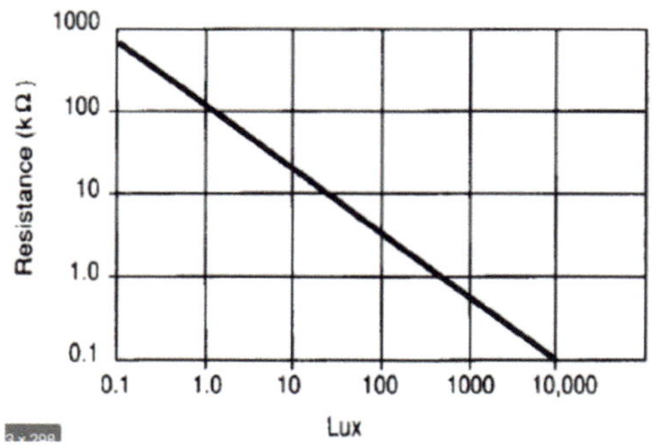

Figure 5.22: Typical LDR characteristics.

Circuit diagram: The circuit diagram of the project is shown in Figure 5.23. A 10 k-ohms fixed resistor is used in the resistive potential divider circuit. The voltage across this resistor is measured using channel 0 of the ADC.

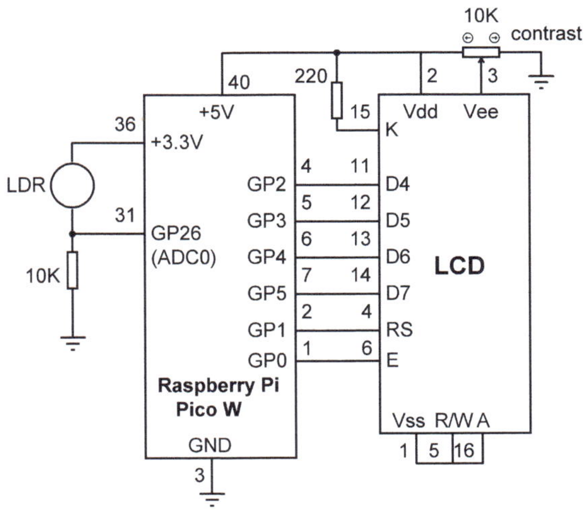

Figure 5.23: Circuit diagram of the project.

Program listing: Figure 5.24 shows the program listing (Program: **LDR**). The program reads the analog voltage across the fixed resistor. Notice that, there is no need to know the absolute voltage. You can just use the digital value of the voltage (0 to 65535) read (**r** in this program). This value should be calibrated to give the intensity of the light in Lux.

```
#----------------------------------------------------------
#              MEASURE THE LIGHT INTENSITY
#              ===========================
#
# In this project an LDR is connected to the Pico in series
# with a fixed resistor. The program displays the voltage
# across the fixed resistor which is proportional to the
# light level falling on the LDR
#
# Author: Dogan Ibrahim
# File  : LDR.py
# Date  : October 2022
#------------------------------------------------------------
from machine import ADC
import utime
import LCDFuncs

LDRin = ADC(0)                          # ADC channel 0
LCDFuncs.lcd_init()                     # Initialize LCD

while True:                             # Do forever
    r = LDRin.read_u16()                # Read LDR
    Tempstr = str(r)[:5]                # Convert to string
    LCDFuncs.lcd_puts(Tempstr)          # Display
```

```
utime.sleep(1)                # Wait 1 second
LCDFuncs.lcd_clear()          # Clear screen
```

Figure 5.24: Program: **LDR**.

Calibration

A light meter is required to calibrate the readings. Measurements should be made at different light levels and a table should be created to list the Lux readings of the meter and the corresponding output from the ADC. Then, a formula can be derived that describes the relationship between the light level and the ADC readings. Alternatively, this table can be indexed for a given ADC reading in order to find the corresponding light level. Interpolation can be made for values between two readings.

5.12 Project 8: Ohmmeter – using parallel LCD

Description: This is an ohmmeter project. The project measures the value of an unknown resistor and displays on the Thonny screen.

Circuit diagram: The circuit diagram of the project is shown in Figure 5.25. A fixed resistor (10 k-ohms) is used in series with the unknown resistor. The program measures the voltage across the fixed resistor and then calculates the value of the unknown resistor (Rx).

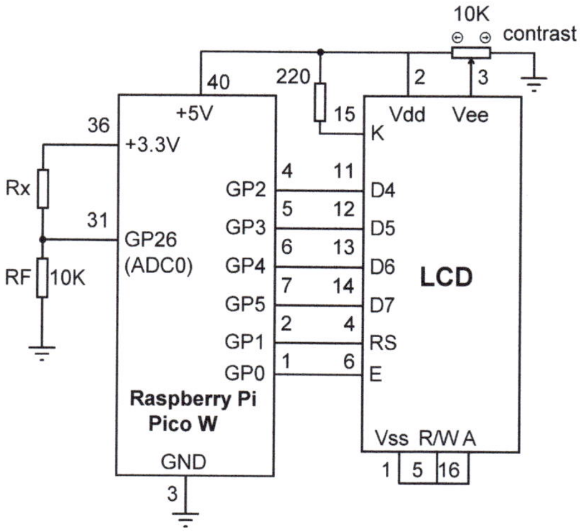

Figure 5.25: Circuit diagram of the project.

Program listing: Figure 5.26 shows the program listing (Program: **Ohmmeter**). If RF is the fixed resistance and Rx is the unknown resistance, assuming the circuit is supplied from +3.3 V (3300 mV), the voltage at the output of the fixed resistor will be:

$$V = 3300 \times RF / (RF + Rx)$$

If you choose RF to be 10 k-ohms, then

$$V = 33000 / (10 + Rx)$$

Where V is in millivolts and RF and Rx are in kilo-ohms. As Rx changes from say 1 k-ohms to 1 megohms, the voltage across the fixed resistor will change from:

$$V = 33000 / 11 = 3000 \text{ mV} \rightarrow V = 33000 / 1001 = 33 \text{ mV}$$

You can easily measure these voltages with the ADC. Therefore, the resistance measurement range of your ohmmeter is well below 1 k-ohms and above 1 megohms.

What you really want is to measure resistance Rx. You can write:

$$Rx = 3300 \times RF/V - RF$$

Remembering that the ADC resolution is 65535 steps and RF = 10 k-ohms, you can write:

$$Rx = 655350 / Vm - RF$$

Where Vm is the digital value directly read from the ADC.

The voltage across the fixed resistor is read 5 times and the value is averaged for higher accuracy.

```
#----------------------------------------------------------------
#                              OHMMETER
#                              ========
#
# In this project the value of an unknown resistor is measured
#
# Author: Dogan Ibrahim
# File  : Ohmmeter.py
# Date  : October 2022
#---------------------------------------------------------------
from machine import ADC
import utime
import LCDFuncs

RF = 10                                 # RF = 10K
LDRin = ADC(0)                          # ADC channel 0
LCDFuncs.lcd_init()                     # Initialize LCD

while True:                             # Do forever
    sum = 0
    for i in range(5):                  # Get 5 readings
```

```
        sum = sum + LDRin.read_u16()    # Read voltage
    Vm = sum / 5                        # Average
    Rx = 65535*RF / Vm - RF             # Calculate Rx
    RxOhms = 1000 * Rx                  # Rx in Ohms
    Tempstr = str(RxOhms)               # Convert to string
    LCDFuncs.lcd_puts(Tempstr)          # Display
    utime.sleep(1)                      # Wait 1 second
    LCDFuncs.lcd_clear()                # Clear screen
```

Figure 5.26: Program: Ohmmeter.

5.13 The I²C LCD

The advantage of using an I²C-based LCD is that only 2 GPIO pins are required to drive the LCD. -based LCD displays are sometimes supplied in two parts: the LCD display, and the I²C controller board. In most standard product distributions, the controller board is soldered to the back of the LCD display as shown in Figure 5.27. The LCD display is basically a 1602-type parallel LCD. The controller board consists of the PCF8574 I²C controller chip (from Texas Instruments or NXP semiconductors), a small potentiometer to adjust the contrast, I/O interface pins, and address selection jumpers.

Figure 5.27: Standard I²C-based LCD.

The controller board has 4 pins: SCL, SDA, VCC, and GND and these must be soldered to the development board I/O pins as shown in Figure 5.28 (if it is not already soldered).

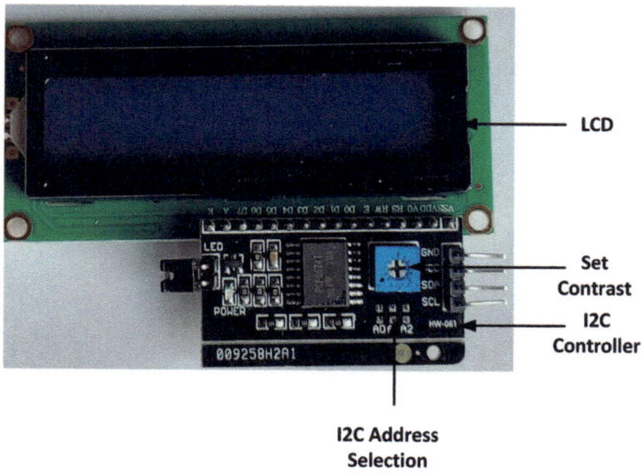

Figure 5.28: Solder the I²C controller board to the LCD.

As shown in Figure 5.29, the I^{2}C address of the PCF8574T chip is selected by 3 jumpers labelled as A0, A1 and A2 on the controller board. By default, the address is set to **0x27** (i.e., no jumper connections).

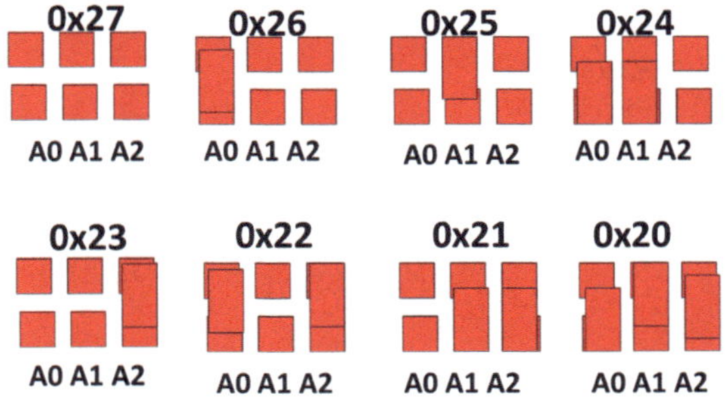

Figure 5.29: address selection.

Before using the I^{2}C LCD, you have to add a I^{2}C library to your Raspberry Pi Pico W. The one used here is from the microcontrollerslab (https://microcontrollerslab.com/i2c-lcd-raspberry-pi-pico-micropython-tutorial/) and it consists of two library files named lcd_api.py and i2c_lcd.py. These library files are available at the book's web site under the program files. Copy these files to your Raspberry Pi Pico with the exact names lcd_api.py and i2c_lcd.py. you can now connect your I^{2}C LCD to the Raspberry Pi Pico W and develop programs.

Hardware interface

The I^{2}C LCD is connected to the Raspberry Pi Pico through the two I^{2}C pins SDA and SCL. I^{2}C LCD displays normally operate with +5 V. However, the Pico is not +5 V compatible. It is safest to use a bi-directional logic level converter module between the display and the Pico. There are many logic converter modules available on the Internet. One such converter module is the TXS0102 (Figure 5.30). This is a two-port logic-level converter module.

Figure 5.30: TXS0102 logic-level converter module.

Figure 5.31 shows the basic connection of the I^2C LCD to the Raspberry Pi Pico W through the logic converter module TXS0102.

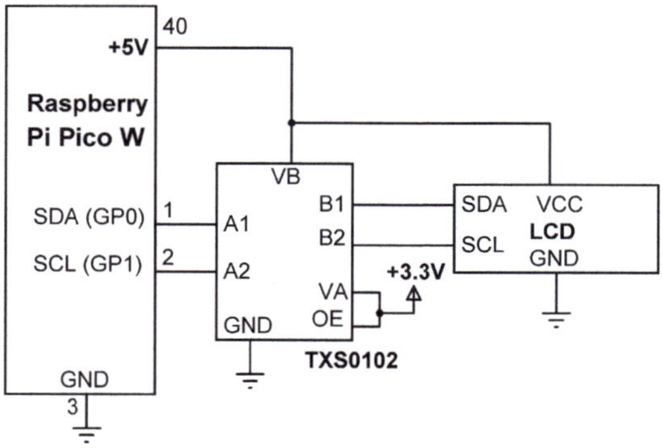

Figure 5.31: Connecting the LCD to the Pico.

Finding the I^2C address of the LCD

The following program will display on the screen the I^2C addresses of all the devices attached to the I^2C bus:

```
import machine
sda=machine.Pin(0)
scl=machine.Pin(1)
i2c=machine.I2C(0,sda=sda, scl=scl, freq=400000)

devices = i2c.scan()
if len(devices) == 0:
    print("No I2C device found")
else:
    print('I2C devices found:', len(devices))
    for device in devices:
        print("I2C address: ",hex(device))
```

A sample run of the above program gave the following display to the author:

I^2C devices found: 1
I^2C address: 0x27

I2C LCD library command:

The I2C library supports the following commands:

clear()
putstr()
backlight_on()
backlight_off()
show_cursor()

hide_cursor()
blink_cursor_on()
blink_cursor_off()
display_on()
display_off()
move_to(column, row) – left hand corner is (0, 0)
putchar()
custom_char()

5.14 Project 9: I²C LCD seconds counter

Description: This is a simple seconds counter project using the I²C LCD.

Aim: The aim of this project is to show how a simple seconds counter can be developed using the I²C LCD.

Block Diagram: Figure 5.32 shows the block diagram.

Figure 5.32: Block diagram of the project.

Circuit diagram: The circuit diagram is the same as in Figure 5.31.

Program listing: Figure 5.33 shows the program listing (Program: **I2CSeconds**). At the beginning of the program, the LCD modules and time module are imported to the program, LCD I²C address, number of rows, and number of columns are defined. Inside the main program, variable count is incremented every second and is displayed on the LCD.

```
#----------------------------------------------------------
#              I2C LCD SECONDS COUNTER
#              =======================
#
# This is a seconds counter project using an I2C LCD. The
# count is displayed on the LCD in the format:
#     Count=nn
#
# Author: Dogan Ibrahim
```

```
# File  : I2CSeconds.py
# Date  : October 2022
#-------------------------------------------------------------
import machine
from machine import I2C
from lcd_api import LcdApi
from i2c_lcd import I2cLcd
import utime

I2C_ADDR = 0x27                        # I2C LCD address
NRows = 2                              # Number of rows
NColumns = 16                          # Number of columns

i2c = I2C(0, sda=machine.Pin(0), scl=machine.Pin(1), freq=400000)
lcd = I2cLcd(i2c, I2C_ADDR, NRows, NColumns)


count = 0
while True:
    lcd.clear()
    count = count + 1
    lcd.putstr("Count=")
    lcd.putstr(str(count))
    utime.sleep(1)
```

Figure 5.33: Program: ***I2CSeconds.***

5.15 Project 10: Internal and external temperature – using LCD

Description: In this project, two temperature sensor chips are used: one to measure the external ambient temperature and the other one to measure the internal ambient temperature. Both readings are displayed on the LCD every 2 seconds.

Block diagram: Figure 5.34 shows the block diagram of the project.

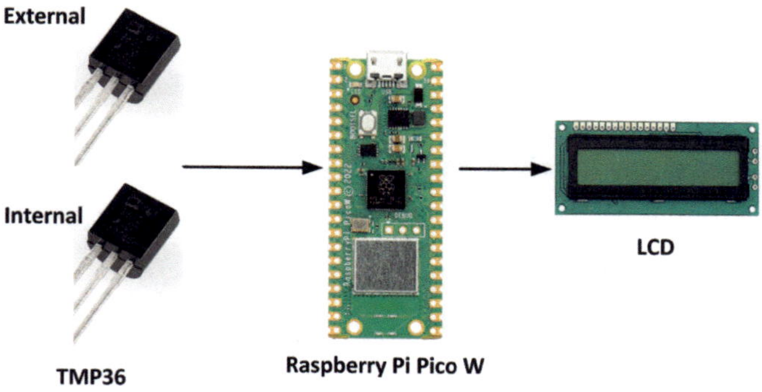

Figure 5.34: Block diagram of the project.

Circuit diagram: The circuit diagram of the project is shown in Figure 5.35. Two TMP36-type temperature sensor chips are used, one connected to channel 0 (external sensor) and the other one connected to channel 1 (internal sensor) of the ADC. The LCD is connected to Pico as in the previous LCD based projects.

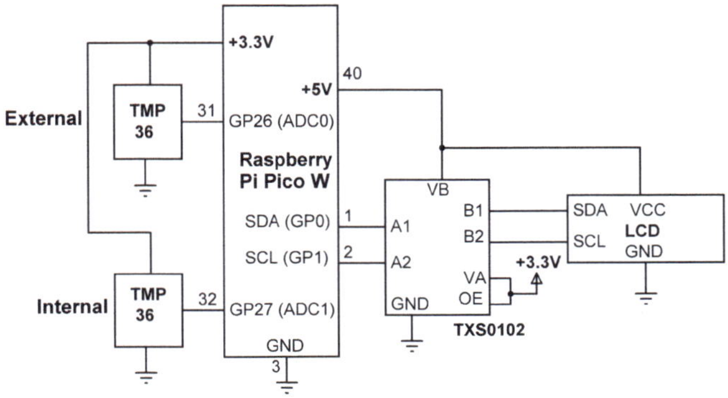

Figure 5.35: Circuit diagram of the project.

Program listing: Figure 5.36 shows the program listing (Program: **MultiTmp**). The external temperature sensor is named **ExtTemp** and is assigned to channel 0. The internal temperature is named **IntTemp** and is assigned to channel 1 of the ADC. Inside the program loop, the temperature of each channel is read, converted into degrees Celsius and displayed on the LCD in the following format (the top row displays the external temperature, the bottom row displays the internal temperature):

Row 0: Ext: nn.mm
Row 1: Int: pp.qq

```
#------------------------------------------------------------
#       EXTERNAL AND INTERNAL TEMPERATURE MEASUREMENT
#       ============================================
#
# This program measures the external and internal temperatures
# using two TMP36 type temperature sensor chips. Both external
# and internal temperatures are displayed on the LCD
#
# Author: Dogan Ibrahim
# File  : MultiTmp.py
# Date  : October 2022
#--------------------------------------------------------------
import machine
from machine import ADC
import utime
from machine import I2C
from lcd_api import LcdApi
from i2c_lcd import I2cLcd
```

```
I2C_ADDR = 0x27                                  # I2C LCD address
NRows = 2                                        # Number of rows
NColumns = 16                                    # Number of columns

i2c = I2C(0, sda=machine.Pin(0), scl=machine.Pin(1), freq=400000)
lcd = I2cLcd(i2c, I2C_ADDR, NRows, NColumns)

ExtTemp = ADC(0)                                 # ADC channel 0
IntTemp = ADC(1)                                 # ADC channel 1
Conv = 3300 / 65535                              # Conversion factor

while True:                                      # Do forever
    Vext = ExtTemp.read_u16()                    # Read channel 0
    mV = Vext * Conv                             # Convert to mV
    Tempext = (mV - 500.0) / 10.0                # External temp
    Vint = IntTemp.read_u16()                    # Read channel 1
    mV = Vint * Conv                             # Convert to mV
    Tempint = (mV - 500.0) / 10.0                # Internal temp
    lcd.clear()                                  # Clear screen
    Tempextstr = str(Tempext)[:5]                # Convert to string
    Tempintstr = str(Tempint)[:5]                # Convert to string
    lcd.putstr("Ext: ")                          # Heading
    lcd.putstr(Tempextstr)                       # Display external
    lcd.move_to(0, 1)                            # Cursor at row 1
    lcd.putstr("Int: ")                          # Heading
    lcd.putstr(Tempintstr)                       # Display internal
    utime.sleep(2)                               # Wait 2 seconds
```

Figure 5.36: Program: **MultiTmp**.

5.16 Project 11: Using a thermistor to measure temperature – using an I²C LCD

Description: In this project, you will be reading the ambient temperature every second using the KY-013 temperature sensor (NTC thermistor) module and then display it on the LCD.

Aim: The aim of this project is to show how the temperature of an NTC thermistor temperature sensor can be read and also how to use an LCD in a project.

Background information: In this project, the KY-013 analog output NTC thermistor type temperature sensor module is used. NTC thermistors are semiconductor devices whose resistances are inversely proportional to the temperature. Thus, as the temperature rises their resistance falls and vice-versa. As shown in Figure 5.37, this is a 3-pin module with the following pins:

GND
+V
S (analog output)

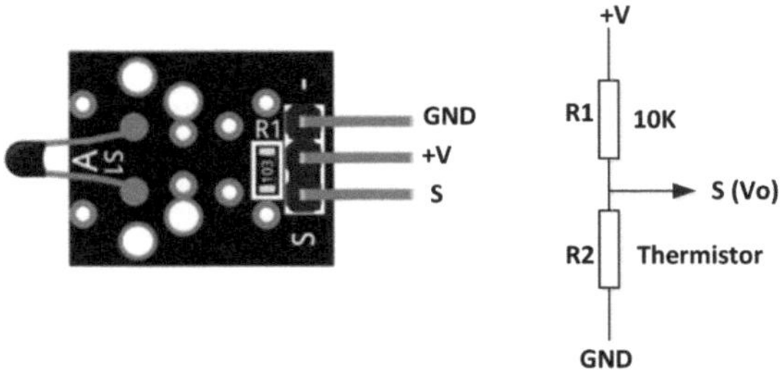

Figure 5.37: KY-013 module.

The sensor module is internally connected to a 10 k-ohms resistor as shown in the figure to form a potential divider circuit. The voltage across the thermistor is read using an analog port of the processor. This voltage is proportional to the temperature where the temperature is calculated using the well-known **Steinhart-Hart** equation. Different thermistors have different **Steinhart-Hart** coefficients, and it is required to know these coefficients in order to calculate the temperature of the thermistor used. For the KY-013, these coefficients are specified by the manufacturer as follows (your coefficients may be slightly different!):

c1 = 0.001129148
c2 = 0.000234125
c3 = 0. 0000000876741

Assuming the resistor is series with the thermistor, R1 = 10 k-ohms (i.e. R1 = 10000 ohms), and a 16-bit ADC is used (0 to 65535 quantization levels) to read the thermistor voltage, the temperature is calculated as follows (see Figure 5.37):

First, calculate the resistance of the thermistor. From the potential divider circuit, the output voltage Vo is given by:

$$Vo = V \times R2 / (R1 + R2) \quad (1)$$

Where V is the applied voltage. From this equation you find R2 as:

$$R2 = Vo \times R1 / (V - Vo) \quad (2)$$

With a 16-bit ADC, if Raw is the raw value read by the ADC, then the actual physical voltage read, Vo, is given by:

$$Vo = Raw \times V / 65535 \quad (3)$$

From (3) and (2) you get:

R2 = [R1 × Raw × V / 65535] / (V – Raw × V/65535] (4)

Equation (4) is simplified to give:

R2 = R1 / (65535/Raw – 1) (5)

or,

R2 = 10000 / (65535 / Raw – 1) (6)

Knowing R2, the temperature is then given by the Steinhart-Hart equation:

T = log(R2) (7)

Tmp = 1 / (c1 + (c2 + (c3 × T × T)) × T) (8)

Now, you convert the temperature from Kelvin into Celsius:

Temp = Tmp – 273.15 (9)

You will be using the above equations to calculate the temperature. Notice that the above calculations are only for 10-kOhms resistor. You should check to make sure that you have the correct resistor installed on the KY-013 sensor module.

Notice: the reading is very sensitive to the resistor and also to the thermistor parameters c1, c2, c3.

Block diagram: Figure 5.38 shows the block diagram of the project.

Figure 5.38: Block diagram of the project.

Circuit diagram: The circuit diagram of the project is shown in Figure 5.39. The sensor is connected to channel 0 of the ADC (GP26, pin 31). The LCD is connected as in the previous projects using LCD.

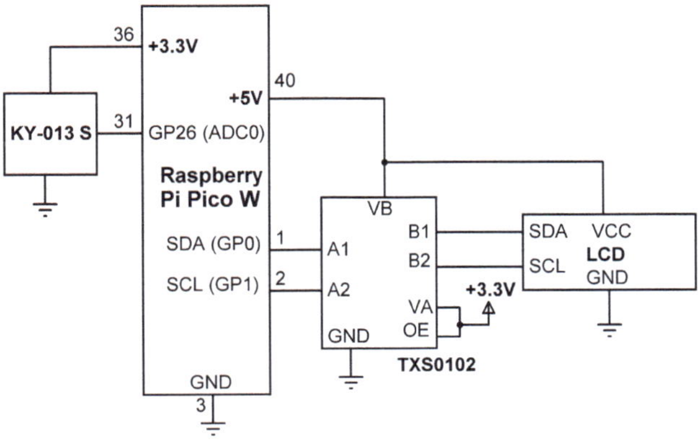

Figure 5.39: Circuit diagram of the project.

Program listing: Figure 5.40 shows the program listing (Program: **Thermistor**). At the beginning of the program, variable Thermistor is assigned to channel 0 of the ADC and the LCD is initialized. The main program runs in a loop every 2 seconds. The thermistor data is read into variable Raw and then function Temperature is called to calculate the temperature. The LCD is cleared, heading Temperature (C) is displayed at the top row, and the temperature reading is formatted and displayed at the bottom row of the LCD in degrees Celsius in the following format:

Row 0:	Temperature (C)
Row 1:	nn.mm

```
#----------------------------------------------------------------
#               THERMISTOR TEMPERATURE MEASUREMENT
#               ==================================
#
# This program measures the ambient temperature using a
# low-cost NTC thermistor module (KY-013). The readings are
# displayed on the LCD
#
# Author: Dogan Ibrahim
# File  : Thermistor.py
# Date  : October 2022
#-----------------------------------------------------------------
import machine
from machine import I2C
from lcd_api import LcdApi
from i2c_lcd import I2cLcd
import utime
from machine import ADC
import math

I2C_ADDR = 0x27                         # I2C LCD address
```

```
NRows = 2                              # Number of rows
NColumns = 16                          # Number of columns

i2c = I2C(0, sda=machine.Pin(0), scl=machine.Pin(1), freq=400000)
lcd = I2cLcd(i2c, I2C_ADDR, NRows, NColumns)

Thermistor = ADC(0)                    # ADC channel 0

#
# Calculate the temperature using Steinhart-Hart equation
#
def Temperature(RawValue):
    c1 = 0.001129148
    c2 = 0.000234125
    c3 = 0.0000000876741
    R1 = 10000.0
    ADC_Res = 65535.0

    R2 = R1 / ((ADC_Res/RawValue - 1))
    T = math.log(R2)
    Tmp = 1.0 / (c1 + (c2 + (c3 * T * T)) * T)
    Temp = Tmp - 273.15
    return Temp

while True:                                  # Do forever
    Raw = Thermistor.read_u16()              # Read channel 0
    temp = Temperature(Raw)                  # Calculate temp
    lcd.clear()                              # Clear LCD
    lcd.putstr("Temperature (C)")            # Display heading
    lcd.move_to(0, 1)                        # Move cursor
    tempstr = str(temp)[:5]                  # Convert to string
    lcd.putstr(tempstr)                      # Display temperature
    utime.sleep(2)                           # Wait 2 seconds
```

Figure 5.40: Program: **Thermistor.**

Note: You could also use the modified form of the Steinhart-Hart equation (known as the **B** parameter equation) if the parameters c1, c2, c3 are not know. The equation of the modified equation is:

$$\mathbf{1/T = 1/T0 + 1/B\ (Log(R/R0))}$$

Where:

T is the measured absolute temperature (take away 273.15 to find the temperature in °C)

T0 is the absolute room temperature (equal to 298.15 K) at 25 °C

B is the thermistor coefficient (usually quoted by the manufacturers, in the region of 3960)

R is the measured resistance of the thermistor

R0 is the resistance of the thermistor at room temperature (usually quoted by the manufacturers, in the region of 10 k-ohms)

5.17 Project 12: Ultrasonic distance measurement – using an I²C LCD

Description: In this project, an ultrasonic sensor module is used to measure the distance to an obstacle in front of the sensor. The distance is displayed on the LCD.

Aim: The aim of this project is to show how the ultrasonic sensor module can be used in a project to measure the distance to an obstacle in front of the sensor.

Ultrasonic sensors

In this project, the popular type HC-SR04 ultrasonic transmitter-receiver module is used (Figure 5.41). The basic features of this sensor module are:

- Operating voltage: 5 V
- Operating current: 2 mA
- Detection distance: 2 cm – 450 cm
- Input trigger signal: 10-µs TTL
- Sensor angle: not more than 15 degrees

Figure 5.41: HC-SR04 ultrasonic sensor module.

The HC-SR04 has the following pin names and descriptions:

Vcc: Power input
Trig: Trigger input
Echo: Echo output
Gnd: Power ground

The basic principle of operation of the HC-SR04 ultrasonic sensor module is as follows (see Figure 5.42):

- A 10-µs trigger pulse is sent to the module.
- The module then sends eight 40-kHz square wave signals to the target and sets the echo pin HIGH.
- The program starts a timer.
- The signal hits the target and echoes back to the module.
- When the signal is returned to the module the echo pin goes LOW.
- The timer is stopped.
- The duration of the echo signal is calculated, and this is proportional to the distance to the target.

The distance to the object is calculated as follows:

Distance to object (in metres) = (duration of echo time in seconds × speed of sound) / 2

The speed of sound is 343 m/s, or 0.0343 cm/µs at 20 °C air temperature

Therefore,

Distance to object (in cm) = (duration of echo time in µs) × 0.0343 / 2

or,

Distance to object (in cm) = (duration of echo time in µs) × 0.0171

For example, if the duration of the echo signal is 294 microseconds, then the distance to the object is calculated as follows:

Distance to object (cm) = 294 × 0.0171 = 5.03 cm

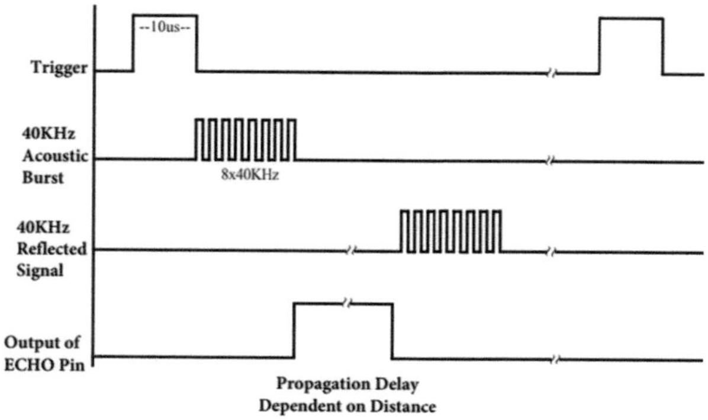

Figure 5.42: Operation of the ultrasonic sensor module.

Block Diagram: Figure 5.43 shows the block diagram of the project.

Figure 5.43: Block diagram of the project.

Circuit Diagram: Figure 5.44 shows the project circuit diagram. Notice that the sensor operates at +5 V and its output is not compatible with the Raspberry Pi Pico W input. A resistive potential divider circuit is used to lower the sensor voltage to +3.3 V.

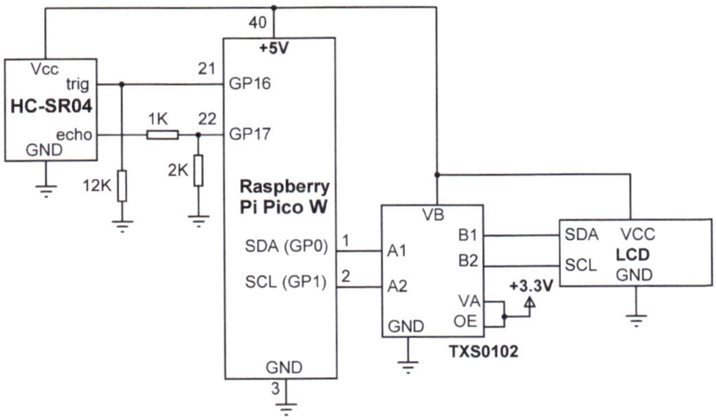

Figure 5.44: Circuit diagram of the project.

Program listing: Figure 5.45 shows the program listing (Program: **Ultrasonic**). At the beginning of the program, trig, and **echo** pins are configured. Inside the main program loop, a trigger pulse is sent for 5 microseconds and the program waits to receive the echo signal (Notice that the required pulse width is 10 microseconds, but the author has measured that using 10 creates a delay much greater than 10 microseconds. A value between 2 and 5 should give correct results). After receiving the echo signal, its duration is calculated and stored in variable Duration. Assuming that the speed of sound in air is 343 m/s (at air temperature 20 ºC), the distance to the object is calculated and stored in variable **distancecm**. This value is then displayed on the LCD as shown in Figure 5.46.

```
#------------------------------------------------------------
#            ULTRASONIC DISTANCE MESUREMENT
#            ==============================
#
# In this project a HC-SR04 type ultrasonic sensor module is
# connected to the Raspberry Pi Pico. The program displays
# distance to an object in-front of the sensor
#
```

```
# Author: Dogan Ibrahim
# File  : Ultrasonic.py
# Date  : October 2022
#-------------------------------------------------------------
import machine
from machine import I2C
from lcd_api import LcdApi
from i2c_lcd import I2cLcd
import utime
from machine import Pin

I2C_ADDR = 0x27                              # I2C LCD address
NRows = 2                                    # Number of rows
NColumns = 16                                # Number of columns

i2c = I2C(0, sda=machine.Pin(0), scl=machine.Pin(1), freq=400000)
lcd = I2cLcd(i2c, I2C_ADDR, NRows, NColumns)

trig = Pin(16, Pin.OUT)                      # trig pin
echo = Pin(17, Pin.IN)                       # echo pin

while True:
    trig.value(0)
    utime.sleep_us(5)                        # Wait until settled

    trig.value(1)                            # Send trig pulse
    utime.sleep_us(5)                        # 5 microseconds
    trig.value(0)                            # Remove trig pulse

    while echo.value() == 0:                 # Wait for echo 1
        pass
    Tmrstrt = utime.ticks_us()

    while echo.value() == 1:                 # Wait for echo 0
        pass
    Tmrend = utime.ticks_us()

    Duration = utime.ticks_diff(Tmrend, Tmrstrt)
    distancecm = Duration * 0.0171
    lcd.clear()
    D = "Dist = " + str(distancecm)[:6] + " cm"
    lcd.putstr(D)
    utime.sleep(1)
```

Figure 5.45: Program: ***Ultrasonic.***

Figure 5.46: Example display on the LCD.

5.18 Project 13: Measuring the depth of a river

Description: The depth of a river is not constant and changes with the amount of precipitation falling in the river, the season, and the tides. Captains of boats or ships on a river need to know the depth of the water in the river before they can sail safely on the river. Also, environmental researchers need to know the depth of a river for analysing the health of the living creatures in the river.

In this project, you will use an HC-SR04-type ultrasonic sensor to measure and display the depth of water in a river every 60 seconds.

Block diagram: Figure 5.47 shows the block diagram of the project. An ultrasonic sensor is mounted at a known height from the top of the river whose depth of water is to be measured. The sensor measures the distance between the sensor and the surface of the water. Assuming the distance between the sensor and the bottom of the river is H, and the distance between the sensor and the surface of the water is h, the depth of water d in the river is given by d = H – h.

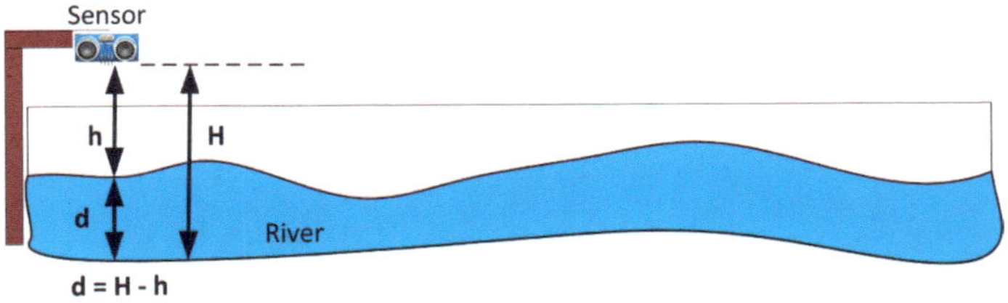

Figure 5.47: Block diagram of the project.

Circuit diagram: The circuit diagram of the project is as in Figure 5.44.

Program listing: Figure 5.48 shows the program listing (Program: **river**). In this project, H is assumed to be 300 cm. the program is similar to the previous one, but here the depth of water is measured and displayed every 60 seconds.

```
#----------------------------------------------------------
#           DEPTH OF WATER IN A RIVER
#           =========================
#
# In this project a HC-SR04 type ultrasonic sensor module is
# connected to the Raspberry Pi Pico W. The program measures
# and displays the depth of a river
#
# Author: Dogan Ibrahim
# File  : river.py
# Date  : October 2022
#------------------------------------------------------------
import machine
from machine import I2C
from lcd_api import LcdApi
from i2c_lcd import I2cLcd
import utime
from machine import Pin

I2C_ADDR = 0x27                       # I2C LCD address
NRows = 2                             # Number of rows
NColumns = 16                         # Number of columns

i2c = I2C(0, sda=machine.Pin(0), scl=machine.Pin(1), freq=400000)
lcd = I2cLcd(i2c, I2C_ADDR, NRows, NColumns)

trig = Pin(16, Pin.OUT)               # trig pin
echo = Pin(17, Pin.IN)                # echo pin
H = 300                               # cm

while True:
    trig.value(0)
    utime.sleep_us(5)                 # Wait until settled

    trig.value(1)                     # Send trig pulse
    utime.sleep_us(5)                 # 5 microseconds
    trig.value(0)                     # Remove trig pulse

    while echo.value() == 0:          # Wait for echo 1
        pass
    Tmrstrt = utime.ticks_us()

    while echo.value() == 1:          # Wait for echo 0
        pass
    Tmrend = utime.ticks_us()
```

```
    Duration = utime.ticks_diff(Tmrend, Tmrstrt)
    h = Duration * 0.0171
    d = H - h
    lcd.clear()
    D = "d = " + str(d)[:7] + " cm"
    lcd.putstr(D)
    utime.sleep(1)
```

Figure 5.48: Program: **river***.*

5.19 Project 14: Ultrasonic reverse parking aid with buzzer

Description: In this project, the ultrasonic sensor module is used together with an active buzzer to help while reverse parking your vehicle. As the distance to the objects gets smaller, the buzzer sounds faster to warn the driver that the objects at the rear of the vehicle are nearer.

Aim: The aim of this project is to show how the ultrasonic sensor module can be used to help reverse park your car.

Block Diagram: Figure 5.49 shows the block diagram of the project.

Figure 5.49: Block diagram of the project.

Circuit diagram: The circuit diagram of the project is shown in Figure 5.50. An active buzzer is added to port pin GP18 of the Raspberry Pi Pico W to sound when the vehicle is close to an obstacle. LCD is not used in this project.

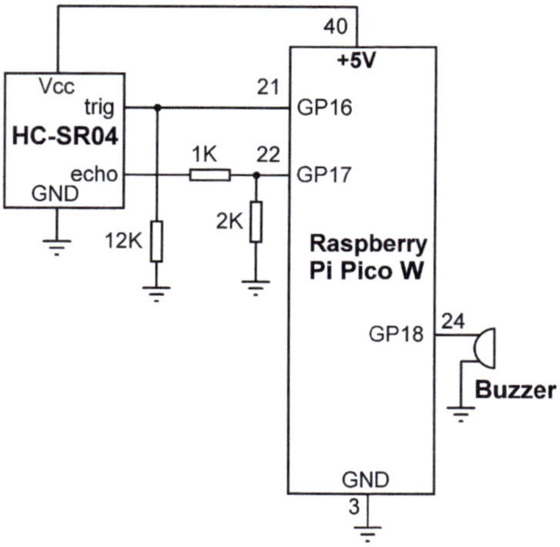

Figure 5.50: Circuit diagram of the project.

Program listing: Figure 5.51 shows the program listing (Program: **Parking**). After finding the distance to the obstacle, the program creates a delay value depending on the distance of the obstacle. As the vehicle gets closer to the obstacle, this delay value is made smaller so that the buzzer sounds faster (more frequently) to alert the driver that the vehicle is getting close to the obstacle. If on the other hand, the vehicle gets further away from the obstacle, then the delay is made larger so that the buzzer sounds slower (less frequently) to inform the driver that the obstacle is not very close.

```
#-----------------------------------------------------------
#           ULTRASONIC REVERSE CAR PARKING AID
#           ==================================
#
# In this project a HC-SR04 type ultrasonic sensor module is
# connected to the Raspberry Pi Pico. Additionally, a buzzer
# is connected. The project sounds the buzzer as the car gets
# near an obstacle.The buzzer sounds faster as the car gets
# nearer an object
#
# Author: Dogan Ibrahim
# File  : Parking.py
# Date  : October 2022
#------------------------------------------------------------
import machine
import utime
from machine import Pin

trig = Pin(16, Pin.OUT)                       # trig pin
echo = Pin(17, Pin.IN)                        # echo pin

Buzzer = Pin(18, Pin.OUT)                     # Buzzer at pin 18
```

```
Buzzer.value(0)                              # Turn OFF buzzer

while True:
    trig.value(0)
    utime.sleep_us(5)                        # Wait until settled

    trig.value(1)                            # Send trig pulse
    utime.sleep_us(5)                        # 5 microseconds
    trig.value(0)                            # Remove trig pulse

    while echo.value() == 0:                 # Wait for echo 1
        pass
    Tmrstrt = utime.ticks_us()

    while echo.value() == 1:                 # Wait for echo 0
        pass
    Tmrend = utime.ticks_us()

    Duration = utime.ticks_diff(Tmrend, Tmrstrt)
    distance = Duration * 0.0171

#
# Now sound the buzzer accordingly.The sounding should be faster
# as the vehicle gets neareer the object.This is done by changing the
# delay in the duration of the sound
#

    if distance > 100:
        dely = 0
    elif distance > 70 and distance < 90:
        dely = 600
    elif distance > 50 and distance < 70:
        dely = 400
    elif distance > 30 and distance < 50:
        dely = 300
    elif distance > 10 and distance < 30:
        dely = 200
    elif distance < 10:
        dely = 10

    if distance < 100:
        Buzzer.value(1)
        utime.sleep_ms(dely)
        Buzzer.value(0)
        utime.sleep_ms(dely)
```

Figure 5.51: Program: ***Parking****.*

5.20 Project 15: Displaying custom characters on the LCD

Description: In this project, you will create an up-arrow shape and display it on your I²C LCD.

Aim: The aim of this project is to show how custom characters can be created and displayed on your LCD.

The block diagram and circuit diagram of the project are as in Figure 5.32 and Figure 5.31 respectively.

Creating custom characters

A custom character consists of 5 columns and 8 rows, i.e., 40 bits. There are many ways of creating a custom character. The web site given below provides an Excel sheet to help create a custom character:

https://peppe8o.com/using-i2c-lcd-display-with-raspberry-pi-pico-and-micropython/

Basically, you have set 1s and 0s to create the character you want. Figure 5.52 shows an up arrow created using the Excel sheet.

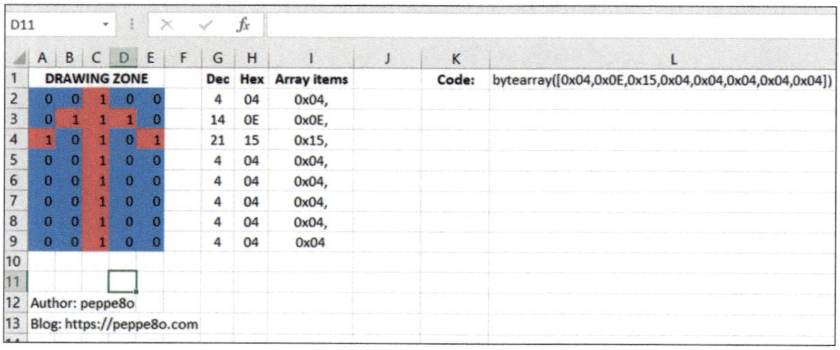

Figure 5.52: Creating an up arrow.

After filling in the required 0s and 1s, look at the code at the right-hand side. In this example the code is:

Code: bytearray([0x04,0x0E,0x15,0x04,0x04,0x04,0x 04,0x04])

You can now write your program.

Program listing: Figure 5.53 shows the program listing (Program: **arrow**). At the beginning of the program, the I²C LCD modules are imported to the program. Then, variable arrow is assigned to the custom code with index 0 (can be 0 to 7). The arrow character is the displayed on the LCD using the **putstr()** function.

```
#------------------------------------------------------------
#               DISPLAY A CUSTOM CHARACTER
#               ==========================
#
# This program displays an arrow character on the LCD
#
# Author: Dogan Ibrahim
# File  : arrow.py
# Date  : October 2022
#--------------------------------------------------------------
import machine
from machine import I2C
from lcd_api import LcdApi
from i2c_lcd import I2cLcd
import utime

I2C_ADDR = 0x27                     # I2C LCD address
NRows = 2                           # Number of rows
NColumns = 16                       # Number of columns

i2c = I2C(0, sda=machine.Pin(0), scl=machine.Pin(1), freq=400000)
lcd = I2cLcd(i2c, I2C_ADDR, NRows, NColumns)

arrow = bytearray([0x04,0x0E,0x15,0x04,0x04,0x04,0x04,0x04])
lcd.custom_char(0, arrow)
lcd.putstr(chr(0))
```

Figure 5.53: Program: ***arrow****.*

Figure 5.54 shows the character displayed on the LCD.

Figure 5.54: Displaying the custom character.

5.21 Project 16: LCD dice

Description: This is an LCD dice project. Pressing a button displays two random dice numbers on the LCD for 3 seconds after which time the display is cleared ready to generate new numbers.

Block diagram: Figure 5.55 shows the block diagram of the project.

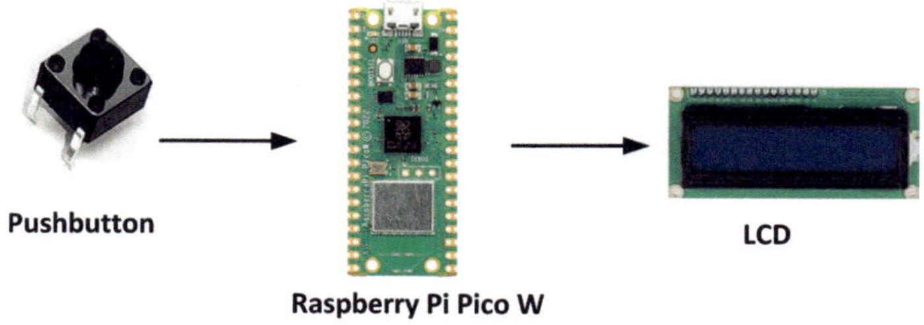

Figure 5.55: Block diagram of the project.

Circuit diagram: The circuit diagram is shown in Figure 5.56. The button is connected to port pin GP17 of the Pico.

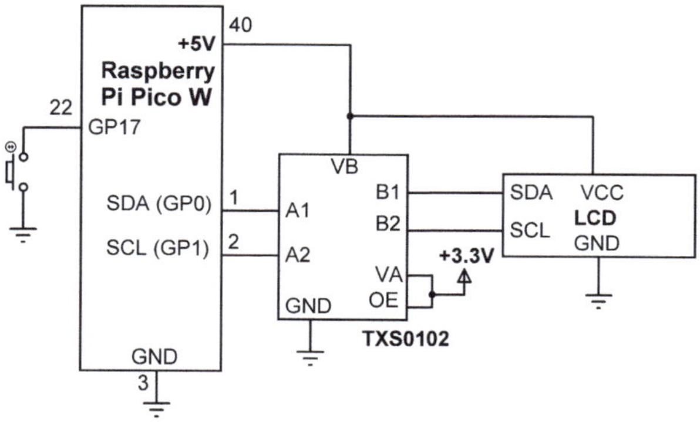

Figure 5.56: Circuit diagram.

Program listing: Figure 5.57 shows the program listing (Program: **LCDDice**). At the beginning of the program, I²C LCD modules are imported to the program. **Button** is configured as an input with internal pull-up resistor enabled. Inside the main program loop, the message **Ready** is displayed and the program waits until the user pressed the button. When the button is pressed, two random numbers are generated between 1 and 6 and are displayed on the LCD for 3 seconds.

```
#------------------------------------------------------------
#               I2C LCD DICE
#               ============
#
# This program displays two dice numbers between 1 and 6
# when a button is pressed
```

```
#
# Author: Dogan Ibrahim
# File  : LCDDice.py
# Date  : October 2022
#-------------------------------------------------------------
import machine
from machine import I2C, Pin
from lcd_api import LcdApi
from i2c_lcd import I2cLcd
import utime
import random

I2C_ADDR = 0x27                    # I2C LCD address
NRows = 2                          # Number of rows
NColumns = 16                      # Number of columns

i2c = I2C(0, sda=machine.Pin(0), scl=machine.Pin(1), freq=400000)
lcd = I2cLcd(i2c, I2C_ADDR, NRows, NColumns)

Button = Pin(17, Pin.IN, Pin.PULL_UP)
lcd.clear()

while True:
    lcd.putstr("Ready")
    while Button.value() == 1:      # Wait for button press
        pass
    lcd.clear()                     # Clear LCD
    rnd1 = random.randint(1, 6)     # Generate a number
    rnd2 = random.randint(1, 6)     # Generate another number
    lcd.putstr(str(rnd1))           # Display first number
    lcd.putstr("    ")
    lcd.putstr(str(rnd2))           # Display second number
    while Button.value() == 0:      # Wait for button release
        pass
    utime.sleep(3)                  # Display for 3 seconds
    lcd.clear()
```

Figure 5.57: Program **LCDDice***.*

5.22 Project 17: Using a real-time clock (RTC) module – setting/displaying date and time

Description: In this project, an RTC module is used to set/get the real time clock for your Raspberry Pi Pico W. The date and time of the RTC module is set from the keyboard and are displayed on I^2C LCD.

RTC Background

The RTC module is an accurate clock module that can be used in microcontroller-based applications. The module provides seconds, minutes, hours, day, date, month, and year information. The leap year is set automatically, where the module is valid until 210.

An RTC module (Figure 5.58) is based on synchronous serial communication (not I²C-compatible). As shown in the figure, the DS1302 RTC chip-based RTC module is used in this project with an on-board crystal for timing. A CR2032 coin-type battery is used to keep the clock running when the module is disconnected from the Pico.

Figure 5.58: The supplied RTC module.

The DS1302 RTC module has the following 4 pins:

VCC:	power supply (+3.3 V)
GND:	GND
CLK:	clock
DAT:	data
RST:	reset

Block diagram: Figure 5.59 shows the block diagram of the project.

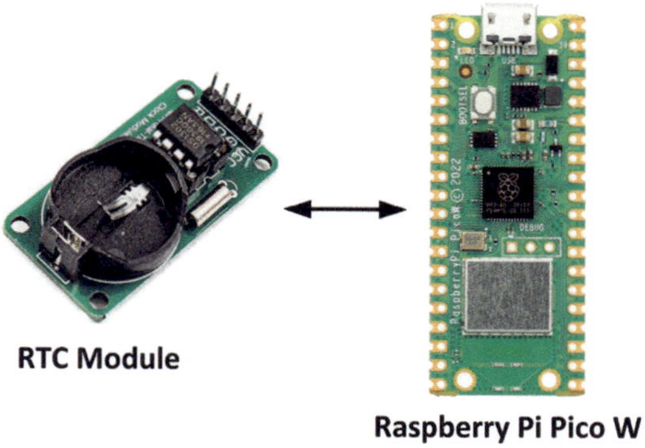

Figure 5.59: Block diagram of the project.

Circuit diagram: The circuit diagram of the project is shown in Figure 5.60. The following connections are made between the RTC module and the Pico:

RTC module	Raspberry Pi Pico W
CLK	GP16
DAT	GP17
RST	GP18
VCC	+3.3V
GND	GND

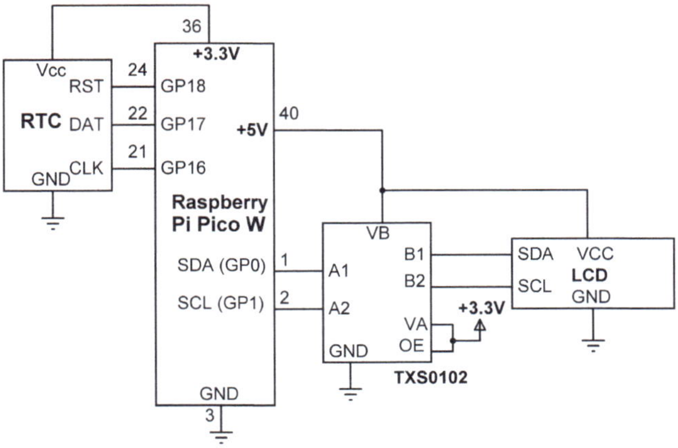

Figure 5.60: Circuit diagram of the project.

Program listing: In this project, the following DS1320 library is used:

https://github.com/omarbenhamid/micropython-ds1302-rtc

You should extract file **ds1302.py** (available on the web site of the book) and then save it on your Raspberry Pi Pico W with the name **ds1302.py** (including **.py**). This library provides the following functions:

start()	start RTC
stop()	stop/pause RTC
date_time(DT = None)	get / set DateTime. If no parameter given, it will return current datetime, otherwise it will set datetime
year(year = None)	get / set year
month(month = None)	get / set month
day(day = None)	get / set day
weekday(weekday = None)	get / set weekday
hour(hour = None)	get / set hour
minute(minute = None)	get / set minute
second(second = None)	get / set second
ram(reg, dat = None)	get / set ram data (31 bytes)

Figure 5.61 shows the program listing (Program: **rtc**). At the beginning of the program, ds1302 and other modules used in the program are imported. The LCD is initialized as in the previous I^2C LCD-based programs. The user is then prompted if the RTC date and/or time will be set. If the answer is **Y** to this question, then the current date and time data are read from the keyboard and then function **date_time()** is called to set the RTC date and time. The date and time information are then displayed on the LCD. Notice that, the RTC is battery-backed and continues to update the date and time even after power is removed from the Pico. If the date and time are correct, the user should enter **N** to the question in order to skip updating the date and/or time.

```
#------------------------------------------------------------
#                               REAL TIME CLOCK MODULE
#                               ======================
#
# This program uses the DS1307 RTC module to set and display
# the date and time on LCD
#
# Author: Dogan Ibrahim
# File  : rtc.py
# Date  : September 2022
#--------------------------------------------------------------
import machine
from machine import I2C, Pin
from lcd_api import LcdApi
from i2c_lcd import I2cLcd
import utime
import ds1302

I2C_ADDR = 0x27                  # I2C LCD address
NRows = 2                        # Number of rows
NColumns = 16                    # Number of columns

i2c = I2C(0, sda=machine.Pin(0), scl=machine.Pin(1), freq=400000)
lcd = I2cLcd(i2c, I2C_ADDR, NRows, NColumns)

ds = ds1302.DS1302(Pin(16),Pin(17),Pin(18))

sel = input("Do you wish to set Date/Time ?: ")
if sel.upper() == "Y":
  print("Date time format is e.g. 2023, 9, 14, 4, 20, 22, 1, 0")
  Y = int(input("Enter Year    : "))
  M = int(input("Enter Month   : "))
  D = int(input("Enter Day     : "))
  day = int(input("Enter day no  : "))
  h = int(input("Enter Hour    : "))
  m = int(input("Enter Minutes : "))
```

```
    s = int(input("Enter Seconds : "))
    ss = int(input("Enter ss       : "))
    ds.date_time([Y,M,D,day,h,m,s,ss])

while True:
    (Y,M,D,day,hr,m,s)=ds.date_time()
    if s < 10:
        s = "0" + str(s)
    if m < 10:
            m = "0" + str(m)
    if hr < 10:
            hr = "0" + str(hr)
    if D < 10:
            D = "0" + str(D)
    if M < 10:
            M = "0" + str(M)
    lcd.move_to(0,0)
    lcd.putstr("Time:")
    lcd.move_to(6,0)
    lcd.putstr(str(hr) + ":" + str(m) + ":" + str(s))
    lcd.move_to(0,1)
    lcd.putstr("Date:")
    lcd.move_to(6,1)
    lcd.putstr(str(D) + "/" + str(M) + "/" + str(Y))
```

Figure 5.61: Program: ***rtc***.

Figure 5.62 shows updating the date and time to current values from the keyboard. An example display is shown in Figure 5.63.

```
Do you wish to set Date/Time ?: Y
Date time format is e.g. 2023, 9, 14, 4, 20, 22, 1, 0
Enter Year    : 2022
Enter Month   : 9
Enter Day     : 15
Enter day no  : 5
Enter Hour    : 16
Enter Minutes : 53
Enter Seconds : 15
Enter ss      : 0
```

Figure 5.62: Updating the date and time.

Figure 5.63: Example display of date and time.

5.23 Project 18: Saving the temperature with time stamping

Description: In this project, the analog temperature sensor chip TMP36 is used together with the RTC module DS1302. Here, the ambient temperature is read every 2 seconds for 10 readings and is stored on the Raspberry Pi Pico W with time stamping. Then, the saved data is displayed on the Thonny screen after 10 seconds.

Aim: The aim of this project is to show how sensor data can be saved on the Pico with time stamping.

Block diagram: Figure 5.64 shows the block diagram of the project.

Figure 5.64: Block diagram of the project.

Circuit diagram: The circuit diagram of the project is shown in Figure 5.65. Here, the TMP36 temperature sensor chip is connected to one of the analog inputs of the Pico. The RTC module is connected to the Pico as in the previous project.

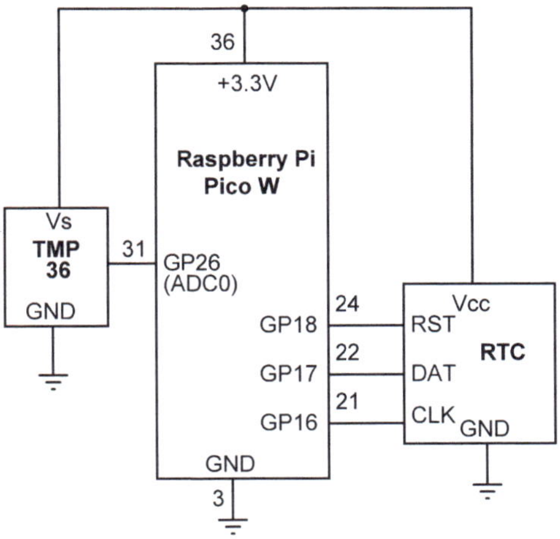

Figure 5.65: Circuit diagram of the project.

Project listing: It is assumed that the current date and time are already set on the RTC module. Figure 5.66 shows the program listing (Program: **rtctmp36**). At the beginning of the program, the modules used are imported and the required number of data samples to collect is stored in variable **samples**. File **Temperature.txt** is opened for writing. A loop is formed to read the ambient temperature every second. The temperature is stored in variable **Temp**. Inside this loop, the date and time are read from the RTC module and saved in variable **Cur**. Then, the temperature readings are stored with a time stamp in the file. At the end of the data collection, the program waits for 10 seconds, opens the file for reading, and reads and displays the collected data.

```
#-----------------------------------------------------------
#              SAVE TEMPERATURE WITH TIME STAMP
#              ================================
#
# This program uses the DS1307 RTC module to read the current
# date and time. The ambient temperature is read and stored on
# the Pico with time stamp. The stored data is then displayed
# on the terminal
#
# Author: Dogan Ibrahim
# File  : rtctmp36.py
# Date  : September 2022
#-------------------------------------------------------------
import machine
from machine import Pin, ADC
import utime
import ds1302

samples = 10
ds = ds1302.DS1302(Pin(16),Pin(17),Pin(18))
```

```
AnalogIn = ADC(0)                                  # ADC channel 0
Conv = 3300 / 65535                                # Conversion factor
fp = open("Temperature.txt", "w")

print("Data Collection Started")
print("======================")
print()

for i in range(samples):
    V = AnalogIn.read_u16()                        # Read temp
    mV = V * Conv                                  # Convert to Volts
    Temp = (mV - 500.0) / 10.0                     # Convert to temp
#
# COnfigrue date and time
#
    (Y,M,D,day,hr,m,s) = ds.date_time()
    if s < 10:
        s = "0" + str(s)
    if m < 10:
            m = "0" + str(m)
    if hr < 10:
            hr = "0" + str(hr)
    if D < 10:
            D = "0" + str(D)
    if M < 10:
            M = "0" + str(M)

    Cur=str(D)+"/"+str(M)+"/"+str(Y)+" "+str(hr)+":"+str(m)+":"+str(s)
    lin = Cur + " " + str(Temp)[:5] + "\n"
    fp.write(lin)
    utime.sleep(2)
fp.close()

#
# Wait 10 seconds, read the data and display on terminal
#
utime.sleep(10)
print("End Of Data Collection. Collected Data:")
print("======================================")

fp = open("Temperature.txt", "r")
for i in range(samples):
    lin = fp.readline()
    print(lin)
```

```
fp.close()
print("End of Program")
```

Figure 5.66: Program: ***rtctmp36****.*

Figure 5.67 shows an example output from the program.

```
Data Collection Started
=======================

End Of Data Collection. Collected Data:
=======================================
15/09/2022 20:01:59 23.33

15/09/2022 20:02:01 23.65

15/09/2022 20:02:03 23.65

15/09/2022 20:02:05 24.13

15/09/2022 20:02:07 26.55

15/09/2022 20:02:09 27.84

15/09/2022 20:02:11 27.84

15/09/2022 20:02:13 26.87

15/09/2022 20:02:15 27.19

15/09/2022 20:02:17 26.23

End of Program
```

Figure 5.67: Example output.

5.24 Project 19: GPS – Displaying the geographical coordinates on LCD

There are cases, especially when working mobile where you may want to know the geographical coordinates (e.g., latitude and longitude) of your station. In this project, you use a GPS to read and display the geographical coordinates of your station on an LCD.

Description: GPS receivers receive geographical data from the GPS satellites, and they provide accurate information about the position of the user on Earth. These satellites circle the Earth at an altitude of about 20,000 km and complete two full orbits every day. In order for a receiver to determine its position, it is necessary for the receiver to communicate with at least 3 satellites. Therefore, if the receiver does not have a clear view of the sky, then it may not be possible to determine its position on Earth. In some applications, external antennas are used so that even weak signals can be received from the GPS satellites.

The data sent out from a GPS receiver is in text format and is known as the NMEA Sentences. Each NMEA sentence starts with a $ character and the values in a sentence are separated by commas. Some of the NMEA sentences returned by a GPS receiver are given below:

$GPGLL: This sentence returns the local geographical latitude and longitude

$GPRMC: This sentence returns the local geographical latitude and longitude, speed, track angle, date, time, and magnetic variation.

$GPVTG: This sentence returns true track, magnetic track, and the ground speed.

$GGGA: This sentence returns the local geographical latitude and longitude, time, fix quality, number of satellites being tracked, horizontal dilution of position, altitude, height of geoid, and DGPS data.

$GPGSV: There are 4 sentences with this heading. These sentences return the number of satellites in view, satellite number, elevation, azimuth, and SNR.

In this project, the GPS Click board (www.mikroe.com) is used. This is a small GPS receiver (see Figure 5.68) which is based on the LEA-6S type GPS. This board operates with +3.3 V and provides two types of outputs: I2C or serial output. In this project, the default serial output is used which operates at a 9600 Baud rate. An external dynamic antenna can be attached to the board in order to improve its reception for indoor use or in for use in places where there may not be clear view of the sky.

Figure 5.68: GPS Click board.

Figure 5.69 shows the complete list of the NMEA sentences output from the GPS Click board every second.

```
$GPGLL,5127.3917,N,00003.13141,E,10534.00,A,A*67
$GPRMC,05305.00,A,5127.35909,,0003.13148,E,0.030,,270919,,,A*7E
$GPVTG,,T,,M,0030,N,0.055,K,A*20
$GGGA,105305.00,5127.35909,N,00003.13148,E,1,09,1.18,46.5,M,45.4,M,,*66
$GPSA,A,3,01,32,08,28,18,03,22,14,11,,,,2.12,1.18,1.76*06
$GPGSV,4,1,13,01,7,304,40,03,40,224,31,08,38,165,32,10,05,054,*77
$GPGSV,4,2,13,11,83,217,3,14,39,094,24,17,17,314,22,18,73,091,41*76
$GPGSV,4,3,13,22,63,219,33,24,1,002,,27,05,150,,28,30,284,28*7F
$GPGSV,4,4,13,32,34,063,35*4E
```

Figure 5.69: NMEA sentences output from the GPS Click board.

GPS Click board is a 2x8 pin dual-in-line module and it has the following pin configuration (pin 1 is the top left pin of the module):

1: No connection	16: No connection
2: Reset	15: No connection

3: No connection	14: TX
4: No connection	13: RX
5: No connection	12: SCL
6: No connection	11: SDA
7: +3.3 V	10: No connection
8: GND	9: GND

In serial operation, only the following pins are required: +3.3 V, GND, TX. In this project, an external dynamic antenna is attached to the GPS Click board as it was used indoors.

$GPGLL and $GPGGA are one of the commonly used NMEA sentences. In this project you will be using the $GPGGA to get the latitude and longitude of your place and then display on the LCD. The $GPGGA sentence format is as follows:

$GPGGA, 161229.487, 3723.2475, N, 12158.3416, W, 1, 07, 1.0, 9.0, M, , , , 0000*18

The fields in this sentence can be decoded as follows:

$GPGGA	GGA protocol header
UTC time	161229.487
Latitude	3723.2475 (37 degrees, 23.2475 minutes)
N/S Indicator	N = North, S = South
Longitude	12158.3416 (121 degrees, 58.3416 minutes)
E/W indicator	E = East or W = West
Position Fix Indicator 1	GPS Sentences Position Fix Indicator
Satellites used	Range is 0 to 12
HDOP	Horizontal Dilution of Precision
MSL Altitude	9.0 Meters
Units	Meters
Geoid Separation	Meters
Units	Meters
Age of diff. corr.	Second
Diff. ref. station ID	0000
Checksum	*18
<CR><LF>	End of message termination

Notice that the fields are separated by commas. The validity of the data is shown by the status field. A value greater than 0 in this field indicates that the data is valid.

Block diagram: Figure 5.70 shows the block diagram of the project.

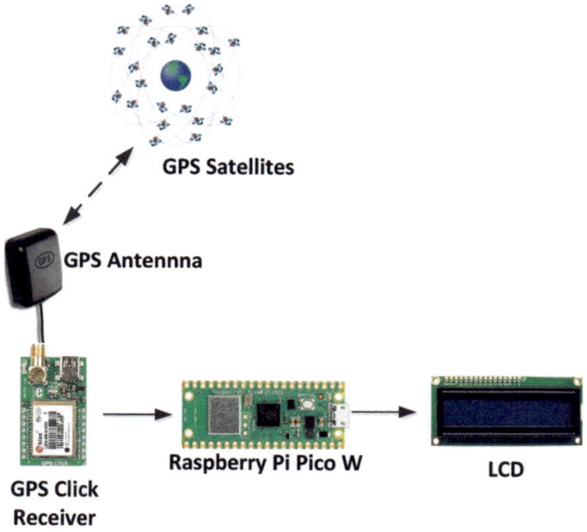

Figure 5.70: Block diagram of the project.

Circuit diagram: The circuit diagram is shown in Figure 5.71. The TX output of the GPS module is connected to port pin GP9 (UART1 RX) of the Pico. The module is operated from +3.3 V. An external antenna is connected to the GPS module since the project was constructed and tested indoors.

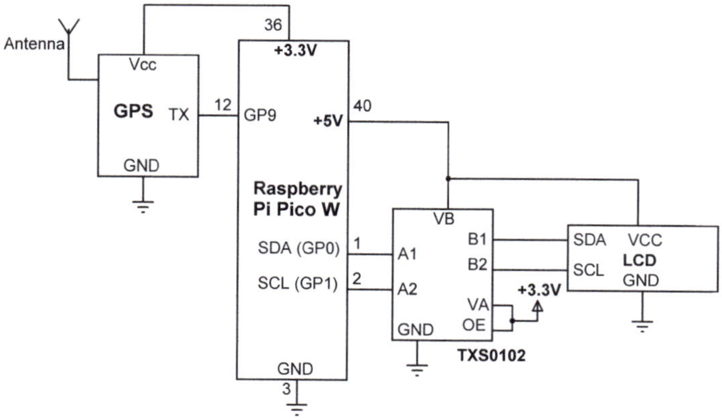

Figure 5.71: Circuit diagram.

Program listing: Figure 5.72 shows the program listing (Program: **gps**). At the beginning of the program, the modules required by the program are imported. Notice that the UART module is also imported since this module is used to receive serial data from the GPS. The UART is initialized to 9600 Baud which is the default communications speed of the GPS module. **tx** and **rx** parameters are set to 8 and 9 which correspond to GP8 and GP9 respectively. These are the UART1 transmit and receive pins of the Pico. Function **Get_GPS()** receives serial data from the GPS module through the UART using function **readline()**. Received serial data is stored in a byte buffer called **buff**. The program looks for the keyword **$GPGGA** in the received data. Once this keyword is detected, function **split()** is used to split the received data into several parts separated by commas. The program then extracts the latitude and longitude together with their directions and returns this data to the main program where they are displayed on the LCD. If the data received from the GPS is not valid, **NO DATA** is displayed on the LCD. Notice that the latitude is at field 2, the latitude direction at field 3, the longitude at field 4, and the longitude direction at field 5.

```
#-----------------------------------------------------------
#                              GPS GEOGRAPHICAL COORDINATES
#                              ============================
#
# This is a GPS project. The geographical coordinates of the
# location is received and displayed on I2C LCD
#
# Author: Dogan Ibrahim
# File  : gps.py
# Date  : October 2022
#-------------------------------------------------------------
import machine
from machine import I2C, UART, Pin
from lcd_api import LcdApi
from i2c_lcd import I2cLcd
import utime

uart = UART(1,baudrate=9600,tx=Pin(8),rx=Pin(9),
            bits=8,parity=None,stop=1)

I2C_ADDR = 0x27                   # I2C LCD address
NRows = 2                         # Number of rows
NColumns = 16                     # Number of columns

i2c = I2C(0, sda=machine.Pin(0), scl=machine.Pin(1), freq=400000)
lcd = I2cLcd(i2c, I2C_ADDR, NRows, NColumns)

buff=bytearray(255)


#
# This function receives and extracts the latitude and longitude
# from the NMEA sentence $GPGGA
#
def Get_GPS():
  while True:
       uart.readline()
       buff = str(uart.readline())
       sdata = buff.split(",")                        # SPlit data
```

```
        if(sdata[0] == "b'$GPGGA" and len(sdata) > 10):
          if sdata[6] == "0":                                   # Valid data?
            return ("No", "DATA")
          else:
           lat = sdata[2]                                       # Get latitude
           latdir = sdata[3]                                    # Latitude dir
           lon = sdata[4]                                       # Get longitude
           londir = sdata[5]                                    # Longitude dir
           deg = lat[0:2]
           min = lat[2:]
           latitude = str(deg) + " " + str(min) + "  " + str(latdir)

           deg = lon[0:3]
           min = lon[3:]
           longitude = str(deg) + " " + str(min) + " " + str(londir)
           return (latitude, longitude)
        else:
           utime.sleep(0.1)

#
# Receive the GPS coordinates and display on the LCD
#
try:
  while True:
      lat, lon = Get_GPS()                                          # Decode
      lcd.clear()
      lcd.move_to(0, 0)
      lcd.putstr(lat)                                               # Display lat
      lcd.move_to(0, 1)
      lcd.putstr(lon)                                               # Display lon
      utime.sleep(0.2)

except KeyboardInterrupt:
   uart.flush()
   uart.deinit()
```

Figure 5.72: Program: ***gps.***

Figure 5.73 shows an example display of the GPS coordinates. The LCD screen should refresh when data is received from the GPS.

Figure 5.73: Example display.

Chapter 6 • Pulse Width Modulation (PWM)

6.1 Overview

Pulse Width Modulation Digital (PWM, also written as: *pulsewidth modulation*) is commonly used to drive heavy loads such as motors, actuators, heaters, and so on. As you shall see in this Chapter, PWM is basically a positive squarewave form whose pulse width and frequency can be changed. By changing the pulse width, you can effectively change the average value of the voltage supplied to the load.

6.2 Basic theory of the pulse width modulation

Pulse Width Modulation (PWM) is a commonly used technique for controlling the power delivered to analog loads using digital waveforms. Although analog voltages (and currents) can be used to control the delivered power, they have several drawbacks. Controlling large analog loads requires large voltages and currents that cannot easily be obtained using standard analog circuits and DACs. Precision analog circuits can be heavy, large, and expensive and they are also sensitive to noise. By using the PWM technique, the average value of voltage (and current) fed to a load is controlled by switching the supply voltage ON and OFF at a fast rate. The longer the power on time, the higher the average voltage supplied to the load.

Figure 6.1 shows a typical PWM waveform where the signal is basically a repetitive positive square wave pulse, having the period T, ON time T_{ON} and OFF time of T - TON seconds. The minimum and maximum values of the voltage supplied to the load are 0 and VP respectively. The PWM switching frequency is usually set to be high (usually in the order of several kHz) so that it does not affect the load that uses the power. The main advantage of PWM is that the load is operated efficiently since the power loss in the switching device is low. When the switch is ON, there is practically no voltage drop across the switch, and when the switch is OFF, there is no current supplied to the load.

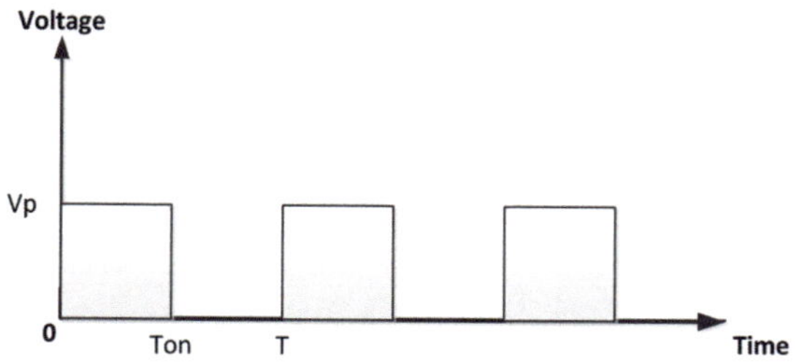

Figure 6.1: PWM waveform.

The duty cycle (or D) of a PWM waveform is defined as the ratio of the ON time to its period. Expressed mathematically,

Duty Cycle (D) = TON / T

The duty cycle is usually expressed as a percentage and therefore,

D = (TON / T) x 100 %

By varying the duty cycle between 0% and 100% you can effectively control the average voltage supplied to the load between 0 and Vp.

The average value of the voltage applied to the load can be calculated by considering a general PWM waveform shown in Figure 1. The average value A of waveform f(t) with period T and peak value ymax and minimum value ymin is calculated as:

$$A = \frac{1}{T}\int_0^T f(t)\,dt$$

or,

$$A = \frac{1}{T}\left(\int_0^{T_{ON}} y_{max}\,dt + \int_{T_{ON}}^{T} y_{min}\,dt \right)$$

In a PWM waveform ymin = 0 and the above equation becomes

$$A = \frac{1}{T}\left(T_{ON}\,y_{max}\right)$$

or

$$A = D\,y_{max}$$

As it can be seen from the above equation, the average value of the voltage supplied to the load is directly proportional to the duty cycle of the PWM waveform and by varying the duty cycle you control the average load voltage. Figure 6.2 shows the average voltage for different values of the duty cycle.

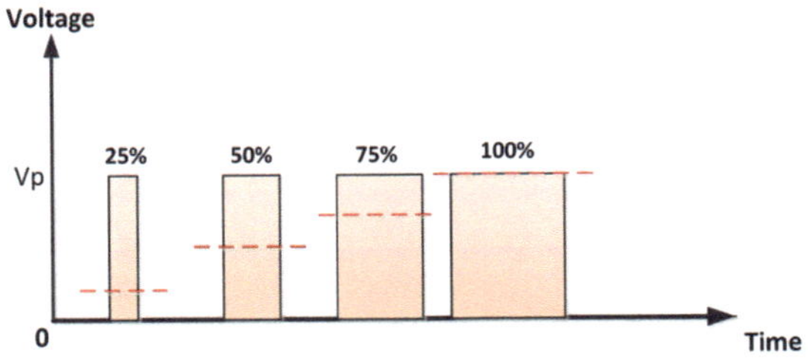

Figure 6.2: Average voltage (shown as dashed line) supplied to a load.

It is interesting to notice that with correct low-pass filtering, the PWM can be used as a DAC if the MCU does not have a DAC channel. By varying the duty cycle, you can effectively vary the average analog voltage supplied to the load.

6.3 PWM channels of the Raspberry Pi Pico W

The Raspberry Pi Pico and Pico W microcontroller have 16 programmable PWM channels. Figure 6.3 shows the pin configurations of these channels. Each channel is identified with a letter and a number, such as **PWM_A[0]**. Some of the 16 PWM channels located on left-hand side of the microcontroller pins are duplicated on the right-hand side and only one of the duplicated channels can be used in an application.

Figure 6.3: PWM channels of the Pico/Pico W.

The PWM channels can be accessed by one of two methods. For example, channel PWM_A[0] connected to port pin GP0 can be accessed as:

```
import machine
ch = machine.PWM(machine.Pin(0))
```

or as

```
from machine import PWM, Pin
ch = PWM(Pin(0))
```

The frequency of the PWM waveform can be set using the following statement. For example, to set the frequency to 1000 Hz:

```
ch.freq(1000)
```

The duty cycle can be set between 0% and 100% by setting it from 0 to 65535. The duty cycle can be set using the following statement. For example, to set the duty cycle to 50%:

ch.duty_u16(32767)

Example projects are given in this Chapter to illustrate how the PWM channels of Pico W can be used.

6.4 Project 1: Generate 1000 Hz PWM waveform with 50% duty cycle

Description: In this project, you create a PWM waveform with a frequency of 1000 Hz and a duty cycle of 50%

Aim: The aim of this project is to show how you can use the Raspberry Pi Pico W PWM functions.

Circuit diagram: In this project port pin GP0 is used and an oscilloscope is connected to this pin to observe the waveform.

Program listing: The program listing is quite simple, and it is given below. Notice that the duty cycle is set to 32767 which corresponds to 50%:

```
from machine import Pin, PWM
ch = PWM(Pin(0))                          # PWM at GP0
ch.freq(1000)                             # Frequency = 1000Hz
ch.duty_u16(32767)                        # 50% duty cycle
while True:
            pass
```

Figure 6.4 shows the generated waveform on the oscilloscope. Here, the horizontal axis was 0.5 ms/division, and the vertical axis was 1 V/division. Clearly, the period of the generated waveform is 1 ms (freq = 1000 Hz), the duty cycle is 50%, and the amplitude is about 3 V.

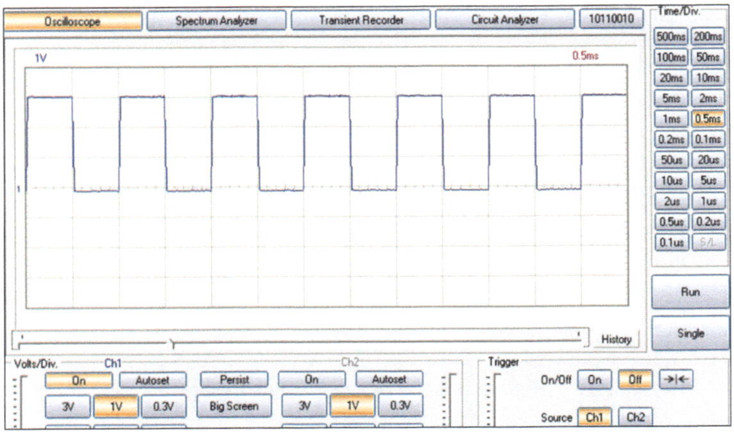

Figure 6.4: Generated PWM waveform.

Note: The author has generated clean and correct PWM waveforms to up to several MHz without any noise.

6.5 Project 2: Changing the brightness of an LED

Description: In this project, an external LED is connected to port pin GP0 through a 470-Ohm current-limiting resistor. The program changes the brightness of the LED by changing the duty cycle of the PWM voltage sent to the LED.

Aim: The aim of this project is to show how the PWM can be used in a project.

The block diagram and the circuit diagram of this project are as in Figure 3.1 and Figure 3.2 respectively where the LED is connected to port pin GP1.

Program listing: Figure 6.5 shows the program listing (Program: **LEDfade**). The frequency is set to 1000 Hz so that the LED light is steady (i.e., not flashing). As the duty cycle is increased from 0% to 100%, the LED brightness increases gradually.

```
#----------------------------------------------------------
#              CHANGING THE BRIGHTNESS OF AN LED
#              =================================
#
# In this program and LED is connected to port pin GP1. The
# brightness of the LED is changed continuously by changing
# duty cycle of the voltage waveform from 0% to 100%
#
# Author: Dogan Ibrahim
# File  : LEDfade.py
# Date  : October 2022
#------------------------------------------------------------
from machine import Pin, PWM
import utime

ch = PWM(Pin(1))                              # PWM at GP1
ch.freq(1000)                                 # Frequency = 1000Hz

i = 0
while True:                                   # Do forever
    ch.duty_u16(i)                            # Change duty cycle
    utime.sleep_ms(300)                       # Delay 300ms
    i = i + 5000                              # Increment i
    if i > 65535:
        i = 0
```

Figure 6.5: Program: ***LEDfade.***

6.6 Project 3: Electronic candle

Description: In the project, an LED is connected to the Pico. The brightness of the LED is changed randomly to give the effect of a candle.

The block diagram and circuit diagram of the project are in Figure 3.1 and Figure 3.2, respectively.

Program listing: Figure 6.6 shows the program listing (Program: **ecandle**). A random number is generated between 0 and 65535 and this number is used to change the duty cycle of the LED, giving the effect of a real candle.

```
#-----------------------------------------------------------
#                              ELECTRONIC CANDLE
#                              =================
#
# In this program an LED is connected to port pin GP1. The
# brightness of the LED is changed randomly to give the
# effect of an electronic candle
#
# Author: Dogan Ibrahim
# File  : ecandle.py
# Date  : October 2022
#-------------------------------------------------------------
from machine import Pin, PWM
import utime
import random

ch = PWM(Pin(1))                       # PWM at GP1
ch.freq(1000)                          # Frequency = 1000Hz

while True:                            # Do forever
    ch.duty_u16(random.randint(100, 65535))
    utime.sleep_ms(300)                # Delay 300ms
```

Figure 6.6: Program: **ecandle**.

6.7 Project 4: Varying the speed of a brushed DC motor

Description: This is a simple project where a small, brushed DC motor is connected to the PICO through a power MOSFET transistor. In addition, a potentiometer is connected to one of the analog inputs of the microcontroller. In this project, the speed of the motor is varied by moving the potentiometer arm.

Block Diagram: Figure 6.7 shows the block diagram of the project. A motor driver (MOSFET transistor) and a potentiometer are connected to the microcontroller.

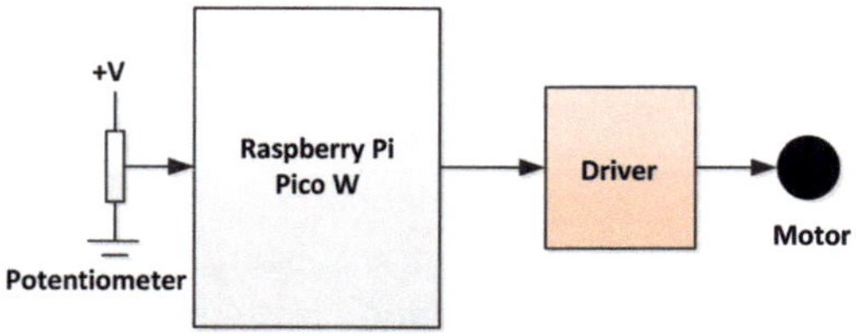

Figure 6.7: Block diagram of the project.

The DC motor in this project is controlled using PWM waves as in the previous project. By varying the potentiometer arm, the analog voltage read by the microcontroller is varied and this in turn changes the PWM duty cycle of the voltage applied to the motor, thus causing the motor speed to change.

Circuit diagram: The circuit diagram of the project is shown in Figure 6.8. The potentiometer is connected to channel 0 of the ADC (GP26, pin 31). The motor is connected to GP17 (pin 22) through an IRL540-type MOSFET switch. It is recommended to use an external power supply for the motor.

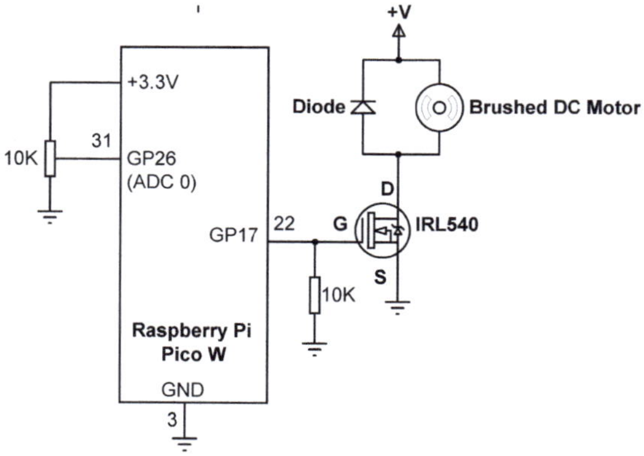

Figure 6.8: Circuit diagram of the project.

Program listing: Figure 6.9 shows the program listing (Program: **Motor**). The data read from the ADC varies between 0 and 65535 as the potentiometer arm is fully moved from one side to the other side. This data is used to change the duty cycle from 0% to 100%.

```
#-------------------------------------------------------------
#            CHANGING THE MOTOR SPEED
#            ========================
#
# In this project a brushed DC motor is connected to the
# Pico W.Additionally, a potentiometer is conencted to channel
# 0 of the ADC.Varying the potentiometer changes the motor speed
```

```
#
# Author: Dogan Ibrahim
# File  : Motor.py
# Date  : October 2022
#--------------------------------------------------------------
from machine import Pin, PWM, ADC

Pot = ADC(0)                        # Pot at channel 0
Motor = Pin(17, Pin.OUT)            # Motor at GP17

ch = PWM(Pin(17))                   # PWM at GP17
ch.freq(1000)                       # Frequency = 1000Hz

while True:                         # Do forever
    duty = Pot.read_u16()           # Read pot data
    ch.duty_u16(duty)               # Change duty cycle
```

Figure 6.9: Program: **Motor**.

6.8 Project 5: Frequency generator with LCD and potentiometer

Description: This is a frequency generator project. A potentiometer and an LCD are connected to the Raspberry Pi Pico W. By varying the potentiometer arm you change the frequency of the generated PWM waveform. The frequency of the generated waveform is displayed on the LCD. The duty cycle of the generated waveform is set at 50%. Frequencies up to about 65535 Hz can be generated by moving the potentiometer arm fully to one side.

Block diagram: Figure 6.10 shows the block diagram of the project.

Figure 6.10: Block diagram of the project.

Circuit diagram: The circuit diagram of the project is shown in Figure 6.11. The potentiometer arm is connected to channel 0 of the ADC (GP0). The LCD is connected as in the previous LCD projects. Output PWM waveform is available at port pin GP16, pin 21).

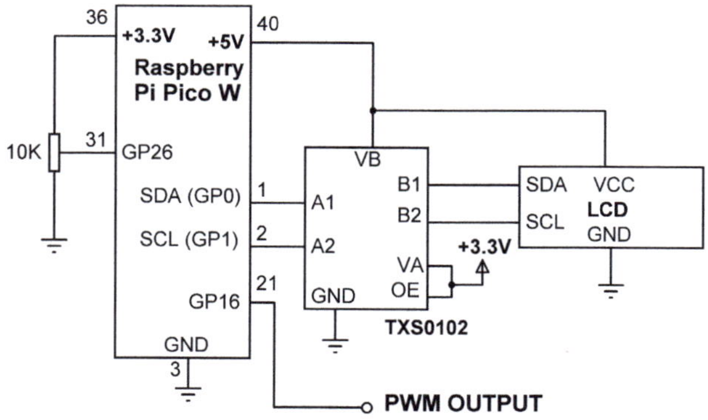

Figure 6.11: Circuit diagram of the project.

Program listing: Figure 6.12 shows the program listing (Program**: FreqGen**). At the beginning of the program, **Pot** is assigned to ADC channel 0, LCD is initialized, and message Frequency(Hz) is displayed at the top row of the LCD. Inside the program loop, the value of **Pot** is read, and this is used to set the frequency of the PWM waveform. The duty cycle is set to 50%.

```
#-----------------------------------------------------------
#           FREQUENCY GENERATOR
#           ===================
#
# In this project a potentiometer is connected to channel 0
# of the Pico. Also, an LCD is connected. The program generates
# PWM waveforms of different frequencies as the potentiometer
# arm is moved
#
# Author: Dogan Ibrahim
# File  : FreqGen.py
# Date  : October 2022
#------------------------------------------------------------
import machine
from machine import I2C
from lcd_api import LcdApi
from i2c_lcd import I2cLcd
import utime
from machine import Pin, PWM, ADC

I2C_ADDR = 0x27                         # I2C LCD address
NRows = 2                               # Number of rows
NColumns = 16                           # Number of columns

i2c = I2C(0, sda=machine.Pin(0), scl=machine.Pin(1), freq=400000)
lcd = I2cLcd(i2c, I2C_ADDR, NRows, NColumns)
```

```
Pot = ADC(0)                          # Pot at channel 0

ch = PWM(Pin(16))                     # PWM at GP16
ch.freq(100)                          # Default freq
lcd.clear()

while True:                           # Do forever
    lcd.move_to(0, 0)
    lcd.putstr("Frequency(Hz)")
    frequency = Pot.read_u16()        # Read pot data
    lcd.move_to(0, 1)
    lcd.putstr(str(frequency))
    ch.duty_u16(32767)
    ch.freq(frequency)                # Change the freuency
    utime.sleep(1)
    lcd.clear()
```

Figure 6.12: Program **FreqGen***.*

The frequency is displayed on the LCD in the following format:

Row 1: Frequency(Hz)
Row 2: nnnnnnn

6.9 Project 6: Measuring the frequency and duty cycle of a PWM waveform

Description: In this project, the frequency and duty cycle of a PWM waveform are read and displayed on the Thonny screen.

Circuit diagram: The PWM waveform whose frequency and duty cycle are to be measured is applied to port GP17 (pin 22). You must make sure that the voltage is not greater than +3.3 V, otherwise, you may damage the input circuitry of your Pico.

Program listing: Figure 6.13 shows the program listing (Program: **MeasFreq**). The program measures the Mark (i.e., logic 1) and Space (i.e., logic 0) timings of the PWM input waveform in microseconds. The duty cycle and the frequency are then calculated as follows:

Duty cycle (%) = 100 × Mark / (Mark + Space)
Frequency (kHz) = 1000 / (Mark + Space)

```
#-----------------------------------------------------------
#               MEASURE THE FREQUENCY AND DUTY CYCLE
#               ====================================
#
# In this project a PWM wave is applied to the PIco. The
# frequency and duty cycle of this wave are measured and
# displayed on the Thonny screen.
#
# Author: Dogan Ibrahim
# File  : MeasFreq.py
# Date  : October 2022
#-------------------------------------------------------------
from machine import Pin
import utime

PWMin = Pin(17, Pin.IN)                     # PWM wave input

while True:                                 # Do forever
    while True:
        if PWMin.value() == 1:              # Wait while 0
            break
    Tmr1Strt = utime.ticks_cpu()            # Start timer 1

    while True:
        if PWMin.value() == 0:              # Wait while 1
            break
    Tmr1End = utime.ticks_cpu()             # End

    while True:
        if PWMin.value() == 1:              # Wait while 0
            break

    Tmr2End = utime.ticks_cpu()             # End

    Mark = utime.ticks_diff(Tmr1End, Tmr1Strt)
    Space = utime.ticks_diff(Tmr2End, Tmr1End)
    duty = 100.0 * Mark / (Mark + Space)
    freqkHz = 1000.0 / (Mark + Space)
    print("Duty Cycle = %5.2f" % duty)
    print("Frequency (kHz) = ", freqkHz, "\n")
    utime.sleep(2)
```

Figure 6.13: Program: ***MeasFreq****.*

Example output from the program is shown in Figure 6.14.

```
Duty Cycle = 39.74
Frequency (kHz) =  1.20048

Duty Cycle = 35.05
Frequency (kHz) =  1.317523

Duty Cycle = 39.59
Frequency (kHz) =  1.203369
```

Figure 6.14: Example output.

6.10 Project 7: Melody maker

Description: This project shows how PWM-type tones with different frequencies can be generated and sent to a passive buzzer device. The project shows how the simple melody **Happy Birthday** can be played on the buzzer.

Aim: The aim of this project is to show how various tones can be generated to create a simple melody.

Block diagram: The block diagram of the project is shown in Figure 6.15.

Figure 6.15: Block diagram of the project.

Circuit diagram: Figure 6.16 shows the circuit diagram of the project. A passive buzzer is connected to port GP0 (pin 1) of the Raspberry Pi Pico W. A transistor switch is used to increase the voltage level of the buzzer (this can be omitted, and the buzzer can be directly connected to GP0 if desired. This, however, will give low output from the buzzer). Any NPN bipolar transistor can be used in this project. The + terminal of the buzzer can be connected to either +3.3 V or to +5 V for higher output from the buzzer.

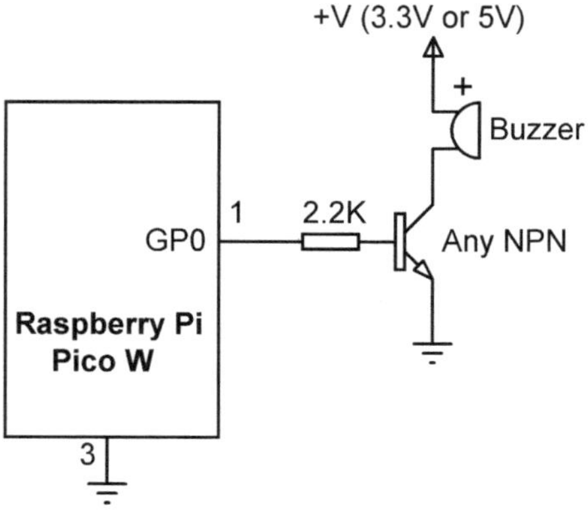

Figure 6.16: Circuit diagram of the project.

Melodies

When playing a melody each note is played for a certain duration and with a certain frequency. In addition, a certain gap is necessary between two successive notes. The frequencies of the musical notes starting from middle C (i.e., C4) are given below. The harmonic of a note is obtained by doubling the frequency. For example, the frequency of C5 is 2 × 262 = 524 Hz.

Notes	C4	C4#	D4	D4#	E4	F4	F4#	G4	G4#	A4	A4#	B4
Hz	261.63	277.18	293.66	311.13	329.63	349.23	370	392	415.3	440	466.16	493.88

To play the tune of a melody, you need to know its musical notes. Each note is played for a certain duration and there is a certain time gap between two successive notes. The next thing you want is to know how to generate a sound with the required frequency and duration. In this project, you will be generating the classic Happy Birthday melody and thus you need to know the notes and their durations. These are given in the table below where the durations are in units of 400 milliseconds (i.e., the values given in the table should be multiplied by 400 to give the actual durations in milliseconds).

Note	C4	C4	D4	C4	F4	E4	C4	C4	D4	C4	G4	F4	C4	C4	C5	A4	F4	E4	D4	A4#	A4#	A4	F4	G4	F4
Duration	1	1	2	2	2	3	1	1	2	2	2	3	1	1	2	2	2	2	2	1	1	2	2	2	4

Program Listing: The program listing (program: **Melody**) is shown in Figure 6.17. The frequencies and durations of the melody are stored in two arrays called **frequency** and **duration,** respectively. Before the main program loop, the durations of each tone are calculated and stored in array **Durations** so that the main program loop does not have to spend any time to do these calculations. Inside the program loop, the melody frequencies are generated with the required durations. Notice that the tone output is stopped by setting the duty cycle to 0. A small delay (100 ms) is introduced between each tone. The melody is repeated after 3 seconds of delay. You can try higher harmonics of the notes for clearer sound. For example, in Figure 6.16 the frequencies are multiplied by 2 to play the second harmonics.

```
#-----------------------------------------------------------------------
#              MELODY MAKER - PLAY HAPPY BIRTHDAY
#              ==================================
#
# In this project a buzzer is connected to port pin GP0
# which is configured as a PWM output. The program plays the
# melody Happy Birthday
#
# Author: Dogan Ibrahim
# File  : Melody.py
# Date  : October 2022
#---------------------------------------------------------------------
from machine import Pin, PWM
import utime

ch = PWM(Pin(0))                                   # PWM output at GP0
MaxNotes = 25
Durations = [0]*MaxNotes
#
# Melody frequencis
#
frequency = [262,262,294,262,349,330,262,262,294,262,
                          392,349,262,262,524,440,349,330,294,466,
                          466,440,349,392,349]

#
# Frequency durations
#
duration = [1,1,2,2,2,3,1,1,2,2,2,3,1,1,2,2,2,2,
                          2,1,1,2,2,2,3]

for k in range(MaxNotes):
    Durations[k] = 400 * duration[k]

while True:                                    # Do forever
    for k in range(MaxNotes):                  # Do for all notes
        ch.duty_u16(32767)                     # Duty cycle
        ch.freq(2*frequency[k])                # Play 2nd harmonics
        utime.sleep_ms(Durations[k])           # Durations
        utime.sleep_ms(100)                    # Wait
    ch.duty_u16(0)                             # Stop playing
    utime.sleep(3)                             # Stop 3 seconds
```

Figure 6.17: Program: ***Melody****.*

Suggestions for additional work

Modify the program given in Figure 6.17 by changing the durations between the notes and see its effects. How can you make the melody run quicker? Also, replace the buzzer with an audio amplifier and a speaker for higher quality and at the same time louder output.

Chapter 7 • TFT Displays

7.1 Overview

A thin-film transistor display is a type of liquid crystal display that makes use of thin-film transistor technology to improve qualities such as contrast and addressability. TFT is also called active-matrix LCD technology. In TFT technology, an individual transistor is used to drive each individual pixel, allowing for much faster response times. The TFT in the LCD controls individual pixels in the display by setting the level of the electric field across the three (red, green, blue) liquid crystal capacitors in the pixel to control the polarization of the crystal material which determines the amount of light that reaches the color filter from the backlight. TFT displays are available in many different physical sizes and by different number pixel numbers.

In this chapter, you will be using a 1.8-inch TFT display in Raspberry Pi Pico W projects.

7.2 TFT display used

The TFT display used in the projects in this Chapter is shown in Figure 7.1. This display has the following specifications:

Size:	1.8 inch
Resolution:	128 × 160
Driver IC:	ST7735
Interface:	4-wire SPI
SD:	On-board SD card adapter

Figure 7.1: TFT display used.

The display has an 8-bit connector on one side for the display logic, and a 4-bit connector on the other side for the SD card interface. The pin definitions are as follows:

8-pin connector	
LED	Display backlight control
SCK	Display SPI Clock
SDA	Display SPI Data
A0	Display address select (data/command, DC)
RESET	Display reset (RST)
CS	Display chip select (CS)
GND	Power supply ground
VCC	Power supply

4-pin connector	
SD-CS	SD card chip select
SD-MOSI	SD card MOSI pin
SD-MISO	SD card MISO pin
SD-SCK	SD card SPI Clock pin

7.3 Connecting the TFT display to Raspberry Pi Pico W

In the projects from this chapter, the TFT display is connected to the Pico as follows (see Figure 7.2), where SPI0 is used for the communication:

TFT display	**Pico port pin**
SCK	GP2 – SPI0 SCK
SDA (or MOSI)	GP3 – SPI0 TX
A0	GP15
RESET	GP14
CS	GP5 - SPI0 CSn
GND	GND
VCC	+3.3V
LED	+3.3V

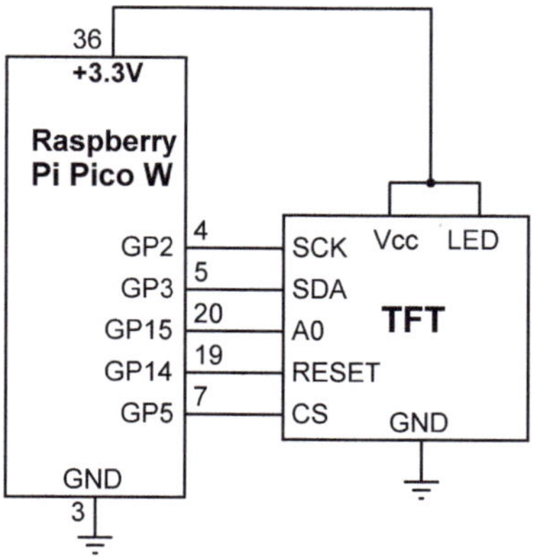

Figure 7.2: TFT display – Pico interface.

7.4 ST7735 TFT display driver library

Before using the 1.8-inch TFT display with your Raspberry Pi Pico, you must load the display driver ST7735 to your Pico. The steps are as follows:

- Copy file **ST7735.py** from the web site of the book to your Raspberry Pi Pico W with the name **ST7735.py**.
- Copy file **sysfont.py** from the web site of the book to your Raspberry Pi Pico W with the name **sysfont.py**.

You can also find files **ST7735.py** and **sysfont.py** in the following **Makerfabs** web site with some other useful files but you will have to change the display size to 128×160, TFTRGB to 0x08, and TFTBGR to 0x00 (see also book: ***Raspberry Pi Pico Experimenting Kit***, sold by Elektor):

https://github.com/Makerfabs/Pico_Primer_Kit/tree/main/example/lib

Figure 7.3 shows the pixel co-ordinates of the TFT display used.

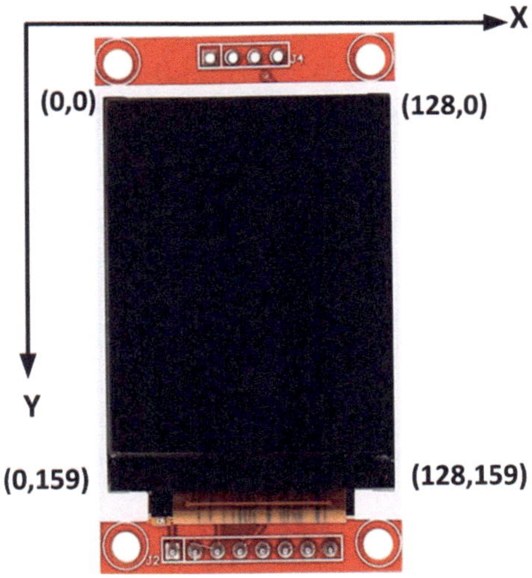

Figure 7.3: Co-ordinates of the display used.

7.4.1 Drawing shapes

Notice that all values are in pixels. Valid colors are:

TFT.BLACK
TFT.RED
TFT.MAROON
TFT.ORANGE
TFT.GOLD
TFT.GREEN
TFT.FOREST
TFT.BLUE

TFT.NAVY
TFT.CYAN
TFT.YELLOW
TFT.PURPLE
TFT.WHITE
TFT.GRAY

Hollow rectangle

Draws a hollow rectangle where the top left coordinates are (x top left, y top left), and the width, height, and color are specified.

rect((x top left, y top left), (width, height), colour)

Filled rectangle

Draws a filled rectangle where the top left coordinates are (x top left, y top left), and the width, height, and colour are specified.

fillrect((x top left, y top left), (width, height), colour)

Hollow circle

Draws a hollow circle where the center x and y coordinates, radius, and the colour are specified

circle((x center, y center), radius, colour)

Filled circle

Draws a filled circle where the center x and y coordinates, radius, and the colour are specified

fillcircle((x center, y center), radius, colour)

Line

Draws a line from the starting point (x start y start) to the end point (x end, y end) and the colour specified

line((x start, y start), (x end, y end), colour)

Vertical line

Draws a vertical line from the starting point (x start y start) and the length and colour specified

vline((x start y start), length, colour)

Horizontal line

Draws a horizontal line from the starting point (x start y start) and the length and colour specified hline((x start, y start), length, colour)

Example 1

Draw two concentric red colour rectangles at the edges of the display with the following co-ordinates:

Hollow red rectangle with top left corner at (0, 0), width = 120, height = 155
Hollow red rectangle with top left corner at (10, 10), width = 110, height = 145

Solution 1

The required program listing is shown in Figure 7.4 (Program: **rectangles)**.

```
#------------------------------------------------------
#              TWO CONCENTRIC RECTANGLES
#              =========================
#
# This program draws two red colour concentric rectangles
#
# Author: Dogan Ibrahim
# File  : rectangles.py
# Date  : October, 2022
#-------------------------------------------------------
from ST7735 import TFT
from sysfont import sysfont
from machine import SPI, Pin

spi = SPI(0, baudrate=20000000, polarity=0, phase=0,
sck=Pin(2), mosi=Pin(3), miso=Pin(4))

tft = TFT(spi, 15, 14, 5)
tft.initg()
tft.rgb(True)

tft.fill(TFT.BLACK)                            # Clear screen

tft.rect((10, 10), (110, 140), TFT.RED)        # Hollow rect
tft.rect((20, 20), (90, 120), TFT.RED)         # Hollow rect
```

Figure 7.4: Program rectangles.

Figure 7.5 shows the display with the two concentric red colour rectangles.

Figure 7.5: The display.

Example 2 – Displaying shapes
Draw the following shapes on the TFT display:

Hollow red rectangle with top left corner at (0, 0), width = 50, height = 50
Filled blue rectangle with top left corner at (60, 0), width = 50, height = 50
Hollow red circle center at (20, 70), with radius = 15
Filled green circle center at (60, 70), with radius = 15
Yellow line from (20, 90) to (60, 100)
Cyan horizontal line starting at (20, 110) and having length 80
Cyan vertical line starting at (110, 60) and having length 70

Solution 2
Figure 7.6 shows the program listing (Program: **TFTBasic**). At the beginning of the program, the required modules are imported to the program and the TFT display SPI interface is defined. Screen background is set to black colour (i.e., screen is cleared) and then the program draws the required shapes.

```
#-----------------------------------------------------
#              VARIOUS SHAPES
#              ==============
#
# This program draws various shapes on the TFT display
#
# Author: Dogan Ibrahim
# File  : TFTBasic.py
# Date  : October, 2022
```

```
#-------------------------------------------------------
from ST7735 import TFT
from sysfont import sysfont
from machine import SPI, Pin

spi = SPI(0, baudrate=20000000, polarity=0, phase=0,
sck=Pin(2), mosi=Pin(3), miso=Pin(4))

tft = TFT(spi, 15, 14, 5)
tft.initg()
tft.rgb(True)

tft.fill(TFT.BLACK)                          # Clear screen

tft.rect((0, 0), (50, 50), TFT.RED)          # Hollow rect
tft.fillrect((60, 0), (50, 50), TFT.BLUE)    # Filled rect
tft.circle((20, 70), 15, TFT.RED)            # Hollow circle
tft.fillcircle((60, 70), 15, TFT.GREEN)      # Filled circle
tft.line((20, 90), (60, 100), TFT.YELLOW)    # Line
tft.hline((20, 110), 80, TFT.CYAN)           # Horiz line
tft.vline((110, 60), 70, TFT.CYAN)           # Vert line
```

Figure 7.6: Program: **TFTBasic**.

Figure 7.7 shows the shapes drawn on the TFT display.

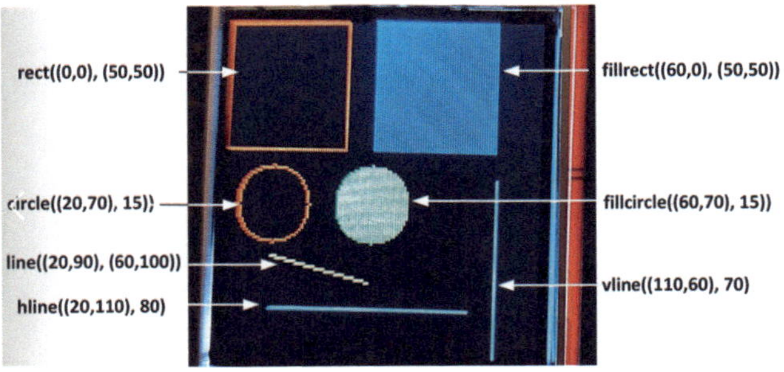

Figure 7.7: Shapes on the TFT display.

7.4.2 Displaying text

Displaying text on the TFT display requires the starting (x, y) coordinate of the text, the text message, colour and font of the text, and additional properties, such as wrapping the text. Function text() is used to display text.

Example 3 - Displaying text

The background is set to white in this example. In this example it is required to display the following text:

Text: PICO **Starting coordinate**: (10,10) **Colour**: Red **Font**: 2 **No wrapping**
Text: PICO **Starting coordinate**: (10,30) **Colour**: Green **Font**: 3 **No wrapping**
Text: PICO **Starting coordinate**: (10, 55) **Colour**: Cyan **Font**: 4 **No wrapping**
Text: PICO **Starting coordinate**: (10, 90) **Colour**: Blue **Font**: 5 **No wrapping**

Solution 3

Figure 7.8 shows the required program listing (Program: **TFTText**).

```
#-------------------------------------------------------
#              DISPLAYING TEXT
#              ===============
#
# This program displays text on the TFT display
#
# Author: Dogan Ibrahim
# File  : TFTText.py
# Date  : October, 2022
#--------------------------------------------------------
from ST7735 import TFT
from sysfont import sysfont
from machine import SPI, Pin

spi = SPI(0, baudrate=20000000, polarity=0, phase=0,
sck=Pin(2), mosi=Pin(3), miso=Pin(4))

tft = TFT(spi, 15, 14, 5)
tft.initg()
tft.rgb(True)

tft.fill(TFT.WHITE)                    # Clear screen
tft.text((10, 10), "PICO", TFT.RED, sysfont, 2, nowrap=True)
tft.text((10, 30), "PICO", TFT.GREEN, sysfont, 3, nowrap=True)
tft.text((10, 55), "PICO", TFT.CYAN, sysfont, 4, nowrap=True)
tft.text((10, 90), "PICO", TFT.BLUE, sysfont, 5, nowrap=True)
```

Figure 7.8: Program: **TFTText***.*

Figure 7.9 shows the texts displayed on the TFT display.

Figure 7.9: Texts on the TFT display.

7.4.3 Other TFT functions

Some other useful TFT functions are:

rotation: You can rotate your drawing. Note that this will *not* rotate what you already drew, but it will change the coordinate system for any new drawing. Rotation function can only take 4 values: 0, 1, 2, and 3. You can only rotate 0º (value 0), 90º (value 1), 180º (value 2) or 270º (value 3) degrees. When rotating, the origin point (0, 0) changes.

invertcolor: Invert the background colour. 1 to set to black, and 0 to set to white

on: turn the display ON (True) or OFF (False)

pixel((x, y), colour): draw a pixel at the given coordinate with the specified colour

fill: fill the background with the specified colour

Example 4 – Other TFT functions

In this example, it is required to perform the following:

Set display background to white
Rotate the display using rotation values 1,2,3,0

Solution 4

Figure 7.10 shows the program listing (Program: TFTOther).

```
#--------------------------------------------------------
#               OTHER TFT FUNCTIONS
#               ===================
#
# This program shows how the rotate function works
#
# Author: Dogan Ibrahim
# File  : TFTOther.py
# Date  : October, 2022
#---------------------------------------------------------
from ST7735 import TFT
from sysfont import sysfont
from machine import SPI, Pin
import utime

spi = SPI(0, baudrate=20000000, polarity=0, phase=0,
sck=Pin(2), mosi=Pin(3), miso=Pin(4))

tft = TFT(spi, 15, 14, 5)
tft.initg()
tft.rgb(True)

tft.fill(TFT.WHITE)                                  # Clear screen

tft.text((10, 30), "PICO", TFT.RED, sysfont, 2, nowrap=True)
utime.sleep(2)

tft.rotation(1)
tft.text((10, 30), "PICO1", TFT.RED, sysfont, 2, nowrap=True)
utime.sleep(2)

tft.rotation(2)
tft.text((10, 50), "PICO2", TFT.RED, sysfont, 2, nowrap=True)
utime.sleep(2)

tft.rotation(3)
tft.text((10, 30), "PICO3", TFT.RED, sysfont, 2, nowrap=True)
utime.sleep(2)

tft.rotation(0)
tft.text((10, 30), "PICO0", TFT.RED, sysfont, 2, nowrap=True)
utime.sleep(2)
```

Figure 7.10: Program: ***TFTOther****.*

Figure 7.11 shows the display when the program is run. Notice that the display shows only the rotated text.

Figure 7.11: Displaying the rotated text.

7.5 Project 1: Seconds counter

Description: This is a seconds counter project. Pressing a button starts the counting. The count is displayed on the TFT display.

Aim: The aim of this project is to show how text and numeric data can be displayed on the TFT display.

Block diagram: Figure 7.12 shows the block diagram of the project.

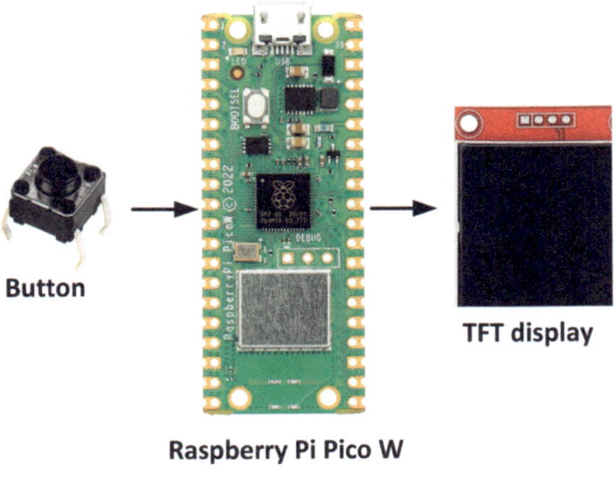

Figure 7.12: Block diagram of the project.

Circuit diagram: The circuit diagram is shown in Figure 7.13 where a button is connected to GP16. The TFT display is connected as in Figure 7.2.

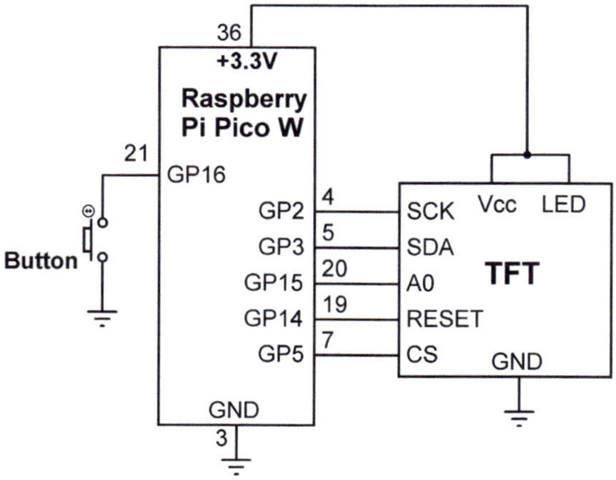

Figure 7.13: Circuit diagram of the project.

Program listing: Figure 7.14 shows the program listing (Program: **TFTCounter**). At the beginning of the program, the modules used are imported to the program. The TFT is initialized and button is assigned to port GP16. Then the program draws a rectangle and displays heading **Seconds Counter** inside this rectangle. The program then waits until the button is pressed. Counting starts as soon as the button is pressed where variable **Count** is incremented every second. The seconds count is displayed on the TFT display.

```
#-----------------------------------------------------
#              SECONDS COUNTER
#              ===============
#
# This is a seconds counter program. The program counts
# up every second. Counting starts when button at GP16
# is pressed
#
# Author: Dogan Ibrahim
# File  : TFTCounter.py
# Date  : October, 2022
#------------------------------------------------------
from ST7735 import TFT
from sysfont import sysfont
from machine import SPI, Pin
import utime

spi = SPI(0, baudrate=20000000, polarity=0, phase=0,
sck=Pin(2), mosi=Pin(3), miso=Pin(4))

Button=Pin(16,Pin.IN, Pin.PULL_UP)       # Button at GP16
```

```
tft = TFT(spi, 15, 14, 5)
tft.initg()
tft.rgb(True)

tft.fill(TFT.BLACK)                       # Clear screen
Count = 0

#
# Display heading
#
tft.rect((10, 10), (110, 100), TFT.WHITE)
tft.text((15, 40), "SECONDS COUNTER", TFT.WHITE, sysfont,
1.1, nowrap=True)

while Button.value() == 1:                # Wait for button
   pass

#
# Increment Count every second and display on the TFT
#
while True:
  tft.text((20, 60), "Count:{:d}".format(Count), TFT.WHITE,
  sysfont, 1, nowrap=True)
  Count = Count + 1
  utime.sleep(1)
```

Figure 7.14: Program: **TFTCounter***.*

Figure 7.15 shows an example display.

Figure 7.15: Example display.

7.6 Project 2: Reaction timer

Description: This is a reaction timer project. An LED is connected to the Pico. This LED turns ON at a random time. The user is required to press a button as soon as he/she sees the LED turned ON. The elapsed time between seeing the LED lit and pressing the button is assumed to be the reaction time and this time is displayed on the TFT display.

Aim: The aim of this project is to show how the elapsed time can be measured, how external interrupt can be set up, and how the reaction time can be displayed on the TFT display.

Block diagram: Figure 7.16 shows the block diagram of the project.

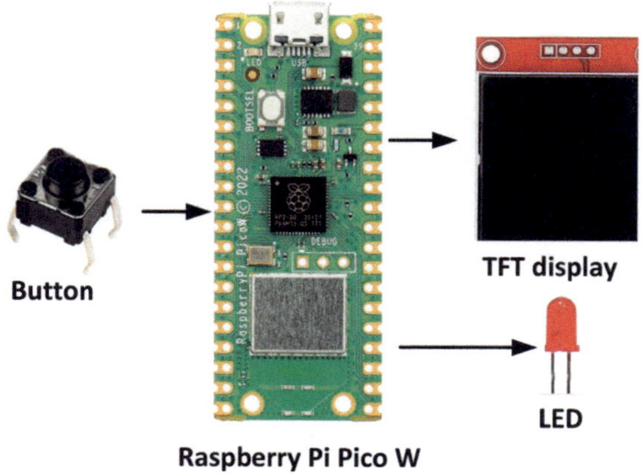

Figure 7.16: Block diagram of the project.

Circuit diagram: The circuit diagram is shown in Figure 7.17. Button and the LED are connected to GP16 and GP17 of PICO, respectively.

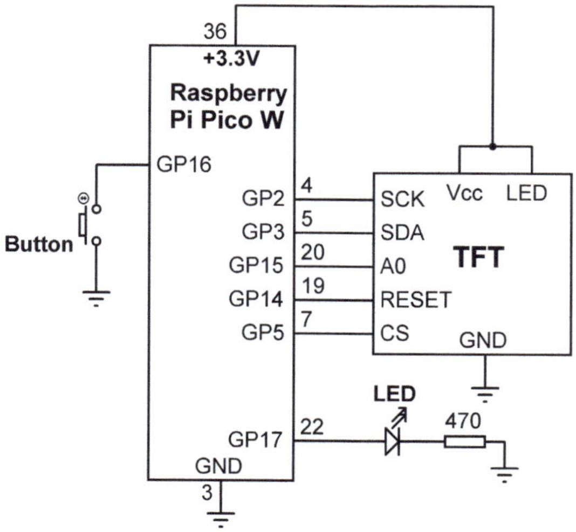

Figure 7.17: Circuit diagram of the project.

Program listing: Figure 7.18 shows the program listing (Program: **TFTReaction**). At the beginning of the program the modules used are imported into the program. The heading REACTION TIMER is displayed on the TFT.

The main program runs in a **while** loop. Inside this loop, the program waits random time, turns ON the yellow LED, and stores the current processor time in variable **TmrStart**. At the same time, external interrupts are enabled so that interrupts can be recognized when the button is pressed. Function **MyButton** is activated when button is pressed. Inside this function, the processor time is read and stored in variable **TmrEnd**. The elapsed time is calculated and variable **ReactionTime** and is then displayed on the TFT. The above process is repeated after 3 seconds of delay.

```
#---------------------------------------------------------------
#                              REACTION TIMER
#                              ==============
#
# This is a reaction timer program which measures the reaction
# of the user and displays on the TFT display in ms. For a
# fast reaction time, the user should press the pushbutton as
# soon as the LED is lit
#
# Author: Dogan Ibrahim
# File  : TFTReaction.py
# Date  : October 2022
#-----------------------------------------------------------------
from ST7735 import TFT
from sysfont import sysfont
from machine import SPI, Pin
import utime
import random

spi = SPI(0, baudrate=20000000, polarity=0, phase=0,
sck=Pin(2), mosi=Pin(3), miso=Pin(4))

#
# Button and LED configuration
#
Button = Pin(16, Pin.IN)                       # Pushbutton
LED = Pin(17, Pin.OUT)                         # LED
flag = 0

tft = TFT(spi, 15, 14, 5)
tft.initg()
tft.rgb(True)
```

```
#
# This is the interrupt service routine. The program jumps here
# as soon as the pushbutton is pressed
#
def MyButton(pin):
    global flag
    Button.irq(handler = None)
    LED.value(1)                                           # LED OFF
    TmrEnd = utime.ticks_ms()                              # End of time
    ReactionTime = utime.ticks_diff(TmrEnd, TmrStart)
    flag = 1
    tft.text((20, 60), "Time:{:.2f} ms".format(ReactionTime), TFT.WHITE,
    sysfont, 1, nowrap=True)
    utime.sleep(3)

#
# Display heading
#
def Heading():
   tft.fill(TFT.BLACK)                                     # Clear screen
   tft.rect((10, 10), (115, 100), TFT.WHITE)
   tft.text((19, 40), "REACTION TIMER", TFT.WHITE, sysfont,
   1.1, nowrap=True)

#
# Start of MAIN program. Generate a random delay and then turn ON LED
#
LED.value(1)                                               # LED OFF

while True:                                                # DO FOREVER
    Heading()
    flag = 0
    rnd = random.randint(3, 10)                            # Random integer number
    utime.sleep(rnd)                                       # Delay random
    LED.value(0)                                           # LED ON
    TmrStart = utime.ticks_ms()
    Button.irq(handler=MyButton, trigger = Pin.IRQ_FALLING)
    while flag == 0:
       pass
```

Figure 7.18: Program: ***TFTReaction****.*

Figure 7.19 shows example display from the program.

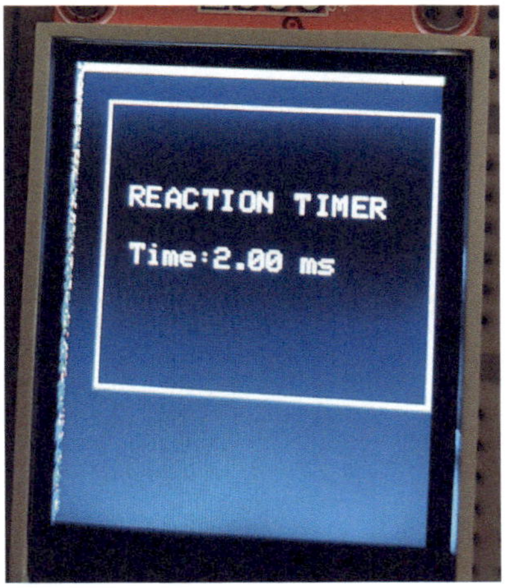

Figure 7.19: Example display.

7.7 Project 3: Temperature and humidity – Display on TFT

Description: This program reads the ambient temperature and humidity from the DHT11 sensor module and displays it on the TFT screen every 5 seconds.

Aim: This program shows how the DHT11 sensor module can be used with the Pico and how the temperature and humidity readings can be displayed on the TFT display.

Block diagram: DHT11 is a low-cost humidity and temperature sensor module (Figure 7.20). A capacitive sensor measures the humidity and a thermistor measures the temperature. The basic specifications of the DHT11 are:

- Low cost
- 3 to 5 V power and I/O
- 2.5 mA max current use during conversion (while requesting data)
- 20-80% humidity readings with 5% accuracy
- 0-50 °C temperature readings ±2 °C accuracy
- No more than 1 Hz sampling rate (once every second)

Figure 7.20: DHT11 temperature and humidity sensor.

Readers may prefer to use the compatible DHT22 which is more accurate than the DHT11

.

Block diagram: Figure 7.21 shows the block diagram.

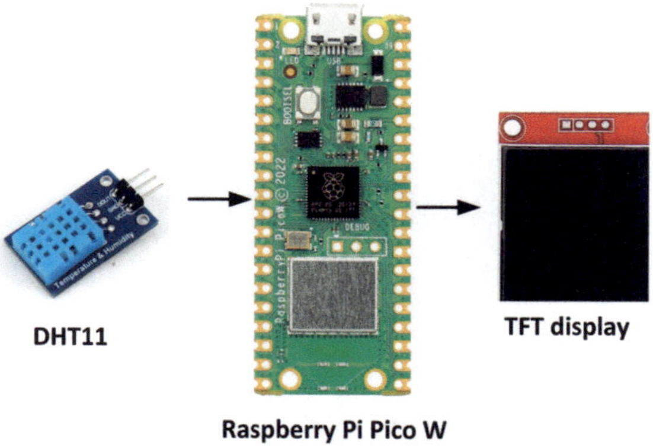

Figure 7.21: Block diagram of the project.

Circuit diagram: Figure 7.22 shows the circuit diagram. DHT11 is connected to GP16.

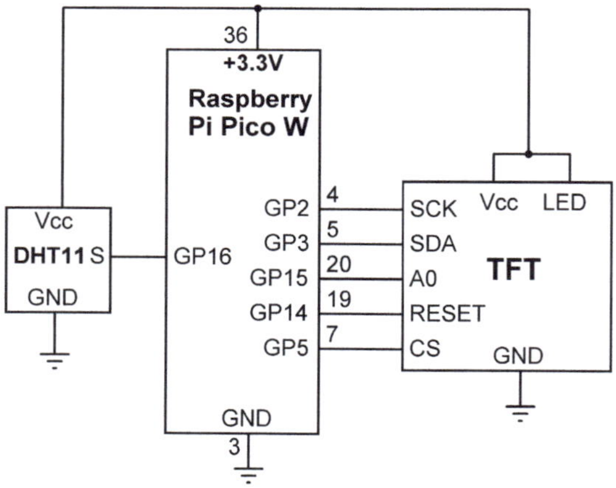

Figure 7.22: Circuit diagram.

Program listing: Before writing the program, you must copy the DHT11/22 driver to your Raspberry Pi Pico W. This is given in the web site of the book with the name **dht.py**. Just copy this file to your Pico with the same name (**dht.py** is also available on the Internet).

Figure 7.23 shows the program listing (Program: **TFTDHT11**). After initializing the SPI bus, the program runs in a loop setup using a while statement. Inside this loop, the heading is displayed, temperature and humidity are read from the DHT11 module and are displayed on the TFT display. This process is repeated after a 5-seconds delay.

```
#---------------------------------------------------------
#              TEMPERATURE AND HUMIDITY ON TFT
#              ===============================
#
# In this program the DHT11 temperature and humidity
# sensor is used and the readings are displayed on TFT
#
# Author: Dogan Ibrahim
# File  : TFTDHT11.py
# Date  : October, 2022
#-----------------------------------------------------------
import utime
from dht import DHT11, InvalidChecksum
from ST7735 import TFT
from sysfont import sysfont
from machine import SPI, Pin

#
# Initialize the SPI
#
spi = SPI(0, baudrate=20000000, polarity=0, phase=0,
sck=Pin(2), mosi=Pin(3), miso=Pin(4))

tft = TFT(spi, 15, 14, 5)
tft.initg()
tft.rgb(True)

#
# Display heading
#
def Heading():
   tft.fill(TFT.WHITE)                                # Clear screen
   tft.rect((5, 10), (100, 120), TFT.RED)
   tft.text((15, 40), "TEMP AND HUM", TFT.BLUE, sysfont,
   1.1, nowrap=True)
   tft.hline((15,51),85, TFT.RED)

#
# Read the ambient temperature and humidity and display on
# the TFT display every 5 seconds
#
while True:
    pin = Pin(16, Pin.OUT, Pin.PULL_DOWN)
    sensor = DHT11(pin)
    t  = (sensor.temperature)
    h = (sensor.humidity)
```

```
Heading()

tft.text((15, 60), "Temp:{:.2f}C".format(t), TFT.BLUE,
sysfont, 1.1, nowrap=True)

tft.text((20, 75), "Hum:{:.2f}%".format(h), TFT.BLUE,
sysfont, 1.1, nowrap=True)

utime.sleep(5)
```

Figure 7.23: Program: **TFTDHT11**.

Figure 7.24 shows an example display of the temperature and humidity on the TFT.

Readers should notice that the DHT11/22 requires precision timing and the program may crash with checksum errors from time to time. You will see in the next project how to capture checksum errors and continue reading the temperature and humidity.

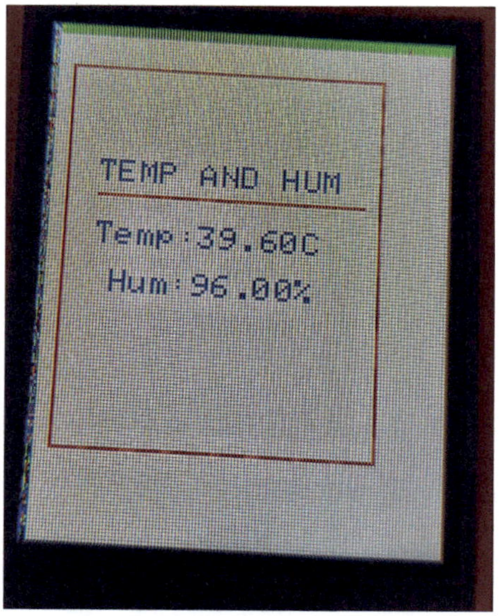

Figure 7.24: Example display.

7.8 Project 4: Minimum/maximum temperature and humidity – Display on TFT

Description: This program is like the previous one, but here minimum and maximum temperatures since the program started are displayed on the TFT display together with the current temperature and humidity.

Aim: This program shows how the DHT11 sensor module can be used with the Pico and how the temperature and humidity readings can be displayed on the TFT display.

The block diagram and circuit diagram of the project are the same as in Figure 7.22 and Figure 7.22.

Program listing: Figure 7.25 shows the program listing (Program: **weather.py**). At the beginning of the program, all the required modules are imported into the program, and LCD is initialized. Inside the main program loop, function **Heading()** is called to display the background boxes where the min, max, current temperature values and humidity will be displayed. The main program then reads the temperature and humidity from DHT11 and displays them in boxes as shown in Figure 7.26. Current temperature is shown in the middle part of the display, humidity at the right-hand side, and minimum and maximum temperatures are displayed inside the bottom left and bottom right-hand side boxes. Note that checksum errors are captured using the try-except statement. The message failed is displayed on the Thonny screen and the program continues after 3 seconds when a checksum error is detected. The program continues after 5 seconds if no errors are detected.

```
#-----------------------------------------------------------
#                              WEATHER ON TFT
#                              ==============
#
# In this program the DHT11 temperature and humidity
# sensor is used and the min and max and current temperature
# and humidity values are displayed on TFT
#
# Author: Dogan Ibrahim
# File  : weather.py
# Date  : October, 2022
#------------------------------------------------------------
import utime
from dht import DHT11, InvalidChecksum
from ST7735 import TFT
from sysfont import sysfont
from machine import SPI, Pin


#
# Initialize the SPI
#
spi = SPI(0, baudrate=20000000, polarity=0, phase=0,
sck=Pin(2), mosi=Pin(3), miso=Pin(4))

tft = TFT(spi, 15, 14, 5)
tft.initg()
tft.rgb(True)


#
# Display heading
#
```

```
def Heading():
    tft.fill(TFT.WHITE)                              # Clear screen
    tft.rect((5, 5), (115, 145), TFT.RED)
    tft.hline((5,115), 113, TFT.RED)
    tft.vline((63, 115), 36, TFT.BLUE)
    tft.vline((75, 5), 112, TFT.BLUE)
    tft.text((85, 10), "HUM", TFT.BLUE, sysfont,
    1.1, nowrap=True)

#
# Read the ambient temperature and humidity and display on
# the TFT display every 5 seconds
#
first = 1

while True:
  try:
    pin = Pin(16, Pin.OUT, Pin.PULL_DOWN)
    sensor = DHT11(pin)                              # Read from DHT11
    t  = (sensor.temperature)                        # Temperature
    h = (sensor.humidity)                            # Humidity
    Heading()                                        # Display Heading

    if first == 1:                                   # min and max
      minT = t
      maxT = t
      first = 0

    if t > maxT:
       maxT = t

    if t < minT:
       minT = t

    tft.text((9, 60), "{:.2f}C".format(t), TFT.MAROON,
    sysfont, 2, nowrap=True)

    tft.text((79, 50), "{:.1f}%".format(h), TFT.BLUE,
    sysfont, 1.1, nowrap=True)

    tft.text((15, 127), "{:.2f}C".format(minT), TFT.BLUE,
    sysfont, 1.1, nowrap=True)

    tft.text((75, 127), "{:.2f}C".format(maxT), TFT.MAROON,
    sysfont, 1.1, nowrap=True)
```

```
    utime.sleep(5)

  except:
    print("failed")
    utime.sleep(3)
```

Figure 7.25: Program: **weather***.*

Figure 7.26: Example display.

7.9 Project 5: ON/OFF temperature control – Setting the desired temperature using buttons and the TFT display

Description: This is an ON/OFF temperature control project where the desired SetTemp is set using three buttons named UP, DOWN, and START. When the program is run, the user is requested to enter the SetPoint temperature. Pressing UP increments the SetPoint. Similarly, pressing DOWN decrements the temperature. When the user is satisfied with the desired SetPoint, he/she must press button START to start the temperature controller.

The aim: The aim of this project is to show how an ON/OFF temperature controller system can be designed using a low-cost temperature sensor module, buttons, and TFT display with the Raspberry Pi Pico W.

Block diagram: The block diagram of the project is shown in Figure 7.27.

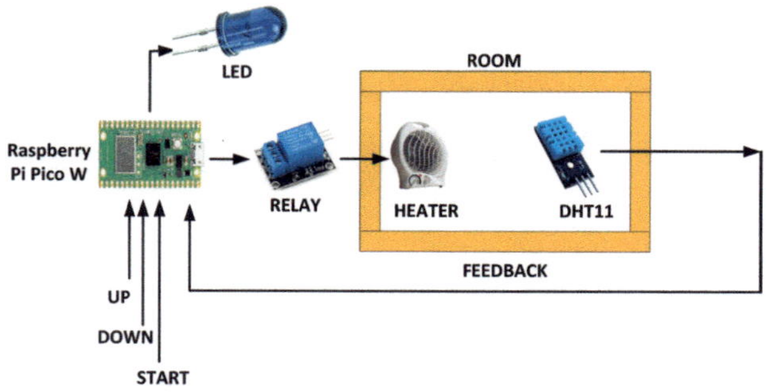

Figure 7.27: Block diagram of the project.

Circuit diagram: Figure 7.28 shows the circuit diagram. The UP, DOWN and START buttons are connected to GP18, GP19 and GP20, respectively. LED and RELAY are connected to GP21 and GP22, respectively.

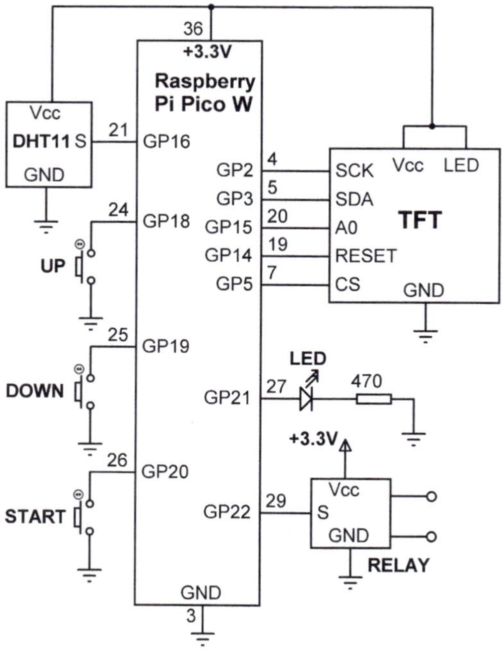

Figure 7.28: Circuit diagram.

Program listing: The program (Program: **ONOFFSET**) listing shown in Figure 7.29 is like the one given in Figure 7.25. Here, additionally, three buttons are configured and used to read the desired temperature **SetTemp**. When the program is run, function **Desired** is called. This function displays a heading and waits until a button is pressed. If **UP** is pressed, the default **SetTemp** is incremented by one, if **DOWN** is pressed, **SetTemp** is decremented by one. If on the other hand button **START** is pressed, then it is assumed that the displayed **SetTemp** is the desired value and the temperature controller action starts. ***Readers should notice that the DHT11/22 is not fully accurate.***

```
#-------------------------------------------------------------
#      ON-OFF TEMPERATURE CONTROL- SETTING THE TEMPERATURE
#      ===================================================
#
# In this program the DHT11 sensor module is used
# together with the relay module and an LED. The room
# temperature is measured every 3 seconds. If the temperature
# is below a set value then the relay and the LED are activated.
# If on the other hand the room temperature is above the set
# value then both the relay and the LED are turned OFF. In this
# program the desired temperature is set using the 3 buttons
#
# Author: Dogan Ibrahim
# File  : ONOFFSET.py
# Date  : October, 2022
#--------------------------------------------------------------
import utime
from dht import DHT11
from ST7735 import TFT
from sysfont import sysfont
from machine import SPI, Pin

#
# Initialize the SPI
#
spi = SPI(0, baudrate=20000000, polarity=0, phase=0,
sck=Pin(2), mosi=Pin(3), miso=Pin(4))

tft = TFT(spi, 15, 14, 5)
tft.initg()
tft.rgb(True)

#
# Set the SetTemp, configure LED, relay, and buttons
#
SetTemp = 10.0                                  # Set point
LED = Pin(21, Pin.OUT)                          # LED
Relay = Pin(22, Pin.OUT)                        # Relay

UP = Pin(18,Pin.IN, Pin.PULL_UP)                # Button
DOWN = Pin(19,Pin.IN, Pin.PULL_UP)              # Button
START = Pin(20,Pin.IN, Pin.PULL_UP)             # Button

LED.value(0)                                    # LED OFF
Relay.value(0)                                  # Relay OFF
t = 10.0
```

```
#
# Display heading
#
def Heading():
   global SetTemp, t
   tft.fill(TFT.WHITE)                             # Clear screen
   tft.rect((5, 10), (110, 100), TFT.RED)
   tft.text((18, 40), "TEMP CONTROL", TFT.BLACK, sysfont,
   1.1, nowrap=True)
   tft.hline((15,51),80, TFT.RED)

   tft.text((15, 60), "SetTemp:{:.2f}C".format(SetTemp), TFT.BLUE,
   sysfont, 1.1, nowrap=True)

   tft.text((15, 75), "   Room:{:.2f}C".format(t), TFT.RED,
   sysfont, 1.1, nowrap=True)

   if Relay.value() == 1:
      tft.text((30, 100), "RELAY ON", TFT.RED, sysfont, 1.1, nowrap=True)
   else:
      tft.text((30, 100), "RELAY OFF", TFT.BLUE, sysfont, 1.1, nowrap=True)

#
# Set the desired temperature (SetTemp). UP increments, DOWN
# decrements, and START starts the temperatue control
#
def Desired():
   global SetTemp
   while START.value() == 1:
      tft.fill(TFT.WHITE)                          # Clear screen
      tft.rect((5, 10), (110, 100), TFT.RED)
      tft.text((15, 30), "SET TEMP", TFT.BLACK, sysfont,
      2, nowrap=True)

      tft.text((15, 80), "SetTemp:{:.2f}C".format(SetTemp), TFT.BLUE,
      sysfont, 1.1, nowrap=True)

      while UP.value() == 1 and DOWN.value() == 1 and START.value() == 1:
         pass
      if UP.value() == 0:
         SetTemp = SetTemp + 1
      elif DOWN.value() == 0:
         SetTemp = SetTemp - 1

Desired()                                              # Read SetTemp
```

```
#
# Read the ambient temperature every 3 seconds and decide
# what to do
#
while True:
    try:
       pin = Pin(16, Pin.OUT, Pin.PULL_DOWN)
       sensor = DHT11(pin)
       t  = (sensor.temperature)                        # Read temp
       if SetTemp > t:                                  # If greater
         Relay.value(1)                                 # Relay ON
         LED.value(1)                                   # ON
       else:
         Relay.value(0)                                 # Relay OFF
         LED.value(0)                                   # LED OFF
       Heading()                                        # Heading
    except:                                             # Error reading
         utime.sleep(10)
         continue
    utime.sleep(10)
```

Figure 7.29: Program: **ONOFFSET**..

Figure 7.30 shows the startup screen when the program is run and the screen when the controller algorithm is running.

Figure 7.30: The startup screen.

7.10 Project 6: ON/OFF temperature control – Setting the desired temperature using rotary encoder and the TFT display

Description: This project is like the previous one but here a rotary encoder is used to set the desired temperature instead of using buttons. Setting the desired temperature with a rotary encoder is fast and easy. Turning the encoder shaft with one click changes the temperature by 1 ºC.

Block diagram: Figure 7.31 shows the block diagram of the project. The temperature settings are displayed on the TFT as the rotary encoder is turned.

Figure 7.31: Block diagram of the project.

Rotary encoder

A rotary encoder (Figure 7.32) is a device that looks like a potentiometer and it senses the rotation and direction of its knob. The device has two internal contacts that make and break a circuit as the knob is turned. As the knob is turned, a click is felt that indicates that the knob has been rotated by one position. With simple logic, you can determine the direction of rotation.

A rotary encoder has the following pins:

GND: power supply ground

Vcc (+): power supply

CLK: This is an output pin used to determine the amount of rotation. Each time the knob is rotated by one click in either direction, the CLK output goes to HIGH and then LOW

DT: This is an output like CLK pin, but it lags the CLK by 90 degrees. This output is used to determine the direction of rotation

SW: This is an active LOW push button. When the knob is pushed, the voltage goes LOW

In your project, each rotation (i.e., click) of the knob will increment (or decrement) the desired temperature by 1 degree. Turning the knob in one direction will increment the count by one, while turning it in the other direction will decrement it by one. When the required value is reached, the user must push the knob so that the program starts to control the temperature.

Figure 7.32: Rotary encoder.

Circuit diagram: The circuit diagram of the project is shown in Figure 7.33. The connections between the Raspberry Pi Pico W and the external components are as follows:

Raspberry Pi Pico W	Pin Number	Connected to
GP2	4	TFT SCK
GP3	5	TFT SDA
GP15	20	TFT A0
GP14	19	TFT RESET
GP5	7	TFT CS
GP16	21	DHT11
GP21	27	LED
GP22	29	RELAY
GP6	9	CLK Rotary encoder
GP7	10	DT Rotary encoder
GP8	11	SW Rotary encoder

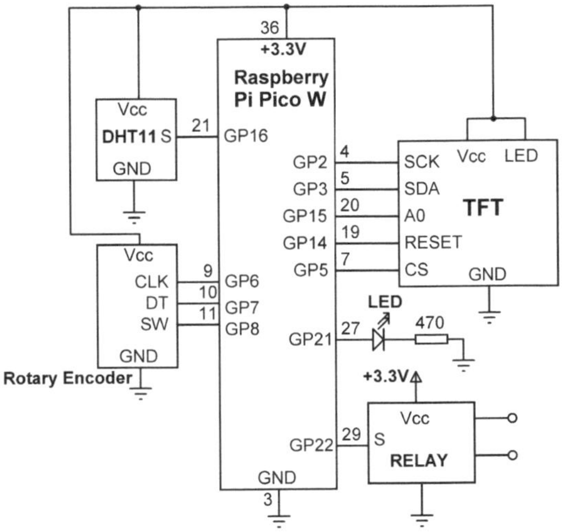

Figure 7.33: Circuit diagram of the project.

Program listing: In this program, you have used the library module **rotary.py**. This module is available at the following web site:

https://github.com/gurgleapps/rotary-encoder

A copy is also available in the web site of the book. This file should be copied to the Pico with the same exact name **rotary.py**.

Figure 7.34 shows the program listing (Program: **TFTRotary.py**). At the beginning of the program, the modules used are imported, LCD connection is defined, LED and RELAY connections are configured, and the RELAY and LED are both turned OFF. Function **Heading()** displays a heading with the **SetTemp**, room temperature, and the RELAY status all displayed and updated every 2 seconds. Function **rotary_changed()** is called whenever the rotary arm is rotated. Rotating clockwise increments **SetTemp** by 1 degree, while rotating it anticlockwise decrements **SetTemp** by 1 degree. The main program reads the room temperature, compares it with the desired temperature and then controls the LED, and the RELAY accordingly.

Notice that the rotary function is interrupt-driven such that the SetTEmp can be changed at any time even when the controller is running.

```
#--------------------------------------------------------------
#      ON-OFF TEMPERATURE CONTROL- SETTING THE TEMPERATURE
#      ===================================================
#
# In this program the DHT11 sensor module is used together
# with a relay module and an LED. The room temperature is
# measured every 3 seconds. If the temperature is below a
# set value then the relay and the LED are activated. If on
# the other hand the room temperature is above the set value
# then both the relay and the LED are turned OFF. In this
# program the desired temperature is set using a rotary encoder
#
# Author: Dogan Ibrahim
# File  : TFTRotary.py
# Date  : October, 2022
#---------------------------------------------------------------
import utime
from dht import DHT11
from ST7735 import TFT
from sysfont import sysfont
from machine import SPI, Pin
from rotary import Rotary
#
# Initialize the SPI
```

```
#
spi = SPI(0, baudrate=20000000, polarity=0, phase=0,
sck=Pin(2), mosi=Pin(3), miso=Pin(4))

tft = TFT(spi, 15, 14, 5)
tft.initg()
tft.rgb(True)

#
# Set the SetTemp, configure LED, and relay
#
global SetTemp
SetTemp = 10.0                                  # Set point
LED = Pin(21, Pin.OUT)                          # LED
Relay = Pin(22, Pin.OUT)                        # Relay

rotary=Rotary(7, 6, 8)                          # DT,CLK,SW

LED.value(0)                                    # LED OFF
Relay.value(0)                                  # Relay OFF

#
# Display heading
#
def Heading():
   global SetTemp, t
   tft.fill(TFT.WHITE)                          # Clear screen
   tft.rect((5, 10), (110, 100), TFT.RED)
   tft.text((18, 40), "TEMP CONTROL", TFT.BLACK, sysfont,
   1.1, nowrap=True)
   tft.hline((15,51),80, TFT.RED)

   tft.text((15, 60), "SetTemp:{:.2f}C".format(SetTemp), TFT.BLUE,
   sysfont, 1.1, nowrap=True)

   tft.text((15, 75), "   Room:{:.2f}C".format(t), TFT.RED,
   sysfont, 1.1, nowrap=True)

   if Relay.value() == 1:
      tft.text((30, 100), "RELAY ON", TFT.RED, sysfont, 1.1, nowrap=True)
   else:
      tft.text((30, 100), "RELAY OFF", TFT.BLUE, sysfont, 1.1, nowrap=True)

#
# Get the desired temperature. Each click of the rotary encoder
# increments or decrements the temperature by 1 degree
```

```
#
def rotary_changed(change):
     global SetTemp
     if change == Rotary.ROT_CW:
         SetTemp = SetTemp + 1
     elif change == Rotary.ROT_CCW:
         SetTemp = SetTemp - 1

rotary.add_handler(rotary_changed)

while True:
    try:
       pin = Pin(16, Pin.OUT, Pin.PULL_DOWN)
       sensor = DHT11(pin)
       t  = (sensor.temperature)                      # Read temp
       if SetTemp > t:                                # If greater
         Relay.value(1)                               # Relay ON
         LED.value(1)                                 # ON
       else:
         Relay.value(0)                               # Relay OFF
         LED.value(0)                                 # BLED OFF
       Heading()                                      # Heading
    except:                                           # Error reading
         utime.sleep(2)
         continue
    utime.sleep(2)
```

Figure 7.34: Program: ***TFTRotary.py****.*

Figure 7.35 shows an example display. Notice that the display is updated every 2 seconds and the **SetTemp** can be changed at any time.

Figure 7.35: Project built on a breadboard.

7.11 Project 7: TFT bitmap display

Description: In this project, small bitmaps are created and displayed on the TFT display. Two bitmaps are created: a heart shape, and an arrow shape. Six such bitmaps are created with spaces between them and displayed across the screen.

Aim: The aim of this project is to show how small bitmaps can be created and displayed on the TFT display.

Program listing: Figure 7.36 shows the program listing (Program: **bitmaps)**. The shapes to be displayed are created using 1s and 0s as shown in the program. Shape **HEART** creates a small heart shape, while shape **ARROW** creates a small arrow shape. The program displays 6 shapes across the screen with 20 pixels of space between each shape. The colors of the shapes can easily be changed by specifying a new color.

```
#-----------------------------------------------------------
#              DISPLAY BITMAP IMAGES
#              ======================
#
# In this program heart images and arrow images are displayed
# across the TFT screen with 20 pixels space between each shape
#
# Author: Dogan Ibrahim
# File  : bitmaps.py
# Date  : October, 2022
#-------------------------------------------------------------
from ST7735 import TFT
from sysfont import sysfont
from machine import SPI, Pin

#
# Initialize the SPI bus for the TFT
#
spi = SPI(0, baudrate=20000000, polarity=0, phase=0,
sck=Pin(2), mosi=Pin(3), miso=Pin(4))

tft = TFT(spi, 15, 14, 5)
tft.initg()
tft.rgb(True)

tft.fill(TFT.BLACK)                          # Clear screen

#
# Heart shape bitmap
#
HEART = [
[ 0, 0, 0, 0, 0, 0, 0, 0, 0],
```

```
[ 0, 1, 1, 0, 0, 0, 1, 1, 0],
[ 1, 1, 1, 1, 0, 1, 1, 1, 1],
[ 1, 1, 1, 1, 1, 1, 1, 1, 1],
[ 1, 1, 1, 1, 1, 1, 1, 1, 1],
[ 0, 1, 1, 1, 1, 1, 1, 1, 0],
[ 0, 0, 1, 1, 1, 1, 1, 0, 0],
[ 0, 0, 0, 1, 1, 1, 0, 0, 0],
[ 0, 0, 0, 0, 1, 0, 0, 0, 0],
]

#
# Arrow shape bitmap
#
ARROW = [
[ 0, 0, 0, 0, 1, 0, 0, 0, 0],
[ 0, 0, 0, 1, 1, 1, 0, 0, 0],
[ 0, 0, 1, 0, 1, 0, 1, 0, 0],
[ 0, 1, 0, 0, 1, 0, 0, 1, 0],
[ 1, 0, 0, 0, 1, 0, 0, 0, 1],
[ 0, 0, 0, 0, 1, 0, 0, 0, 0],
[ 0, 0, 0, 0, 1, 0, 0, 0, 0],
[ 0, 0, 0, 0, 1, 0, 0, 0, 0],
[ 0, 0, 0, 0, 1, 0, 0, 0, 0],
]

#
# Clear and display 6 shapes across the display
#

for i in range(0, 120, 20):
 for y, row in enumerate(HEART):
    for x, c in enumerate(row):
        if c == 1:
           tft.pixel((x+i, y+30), TFT.WHITE)
        else:
           tft.pixel((x+i, y+30), TFT.BLACK)
```

Figure 7.36: Program: ***bitmaps****.*

Figure 7.37 shows the heart shapes and arrow shapes displayed across the screen.

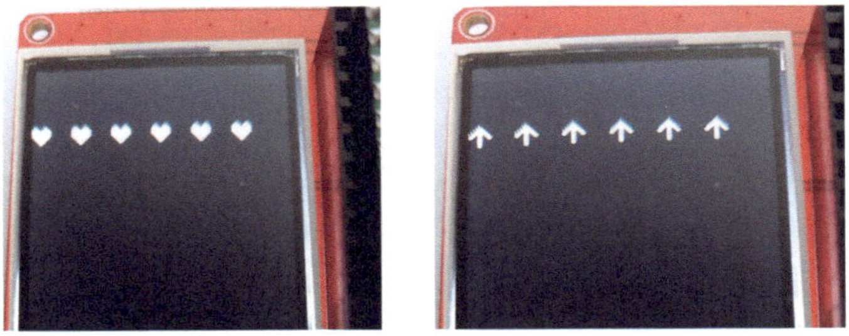

Figure 7.37: Displaying the created shapes.

7.12 Project 8: Using a 4×4 keypad

Description: This is a 4×4 keypad program. The program reads the key pressed by the user and displays its code on the Thonny screen.

Aim: The aim of this project is to show how a 4×4 keypad can be used together with a TFT display in a Raspberry Pi Pico W project.

The 4×4 Keypad: There are several types of keypads that can be used in microcontroller-based projects. In this project a 4×4 keypad (see Figure 7.38) is used. This keypad has keys for numbers 0 to 9 and letters A, B, C, D, *, and #. The keypad is interfaced to the processor with 8 wires with the names R1 to R4 and C1 to C4, representing the rows and columns respectively of the keypad (see Figure 7.39).

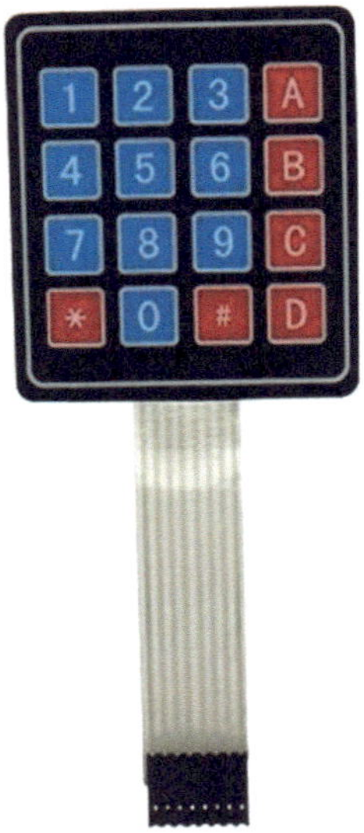

Figure 7.38: 4×4 keypad.

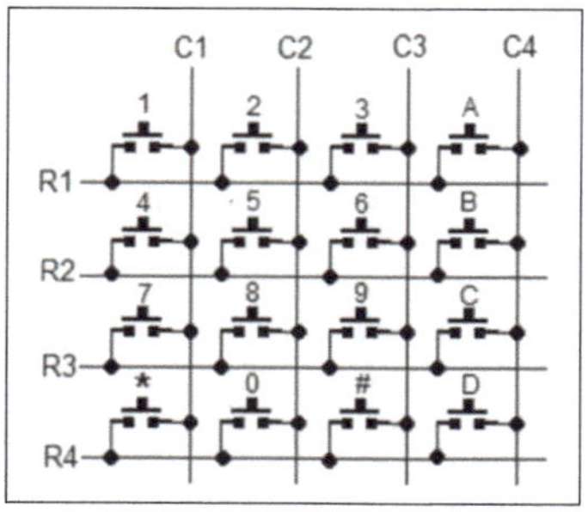

Figure 7.39: Circuit diagram of the 4×4 keypad.

The operation of the keypad is quite simple: the columns are configured as outputs and the rows as inputs. The pressed key is identified by using column scanning. Here, a column is forced low while the other columns are held high. Then the state of each row is scanned, and if a row is found to be low, then the key at the intersection of the row which is low, and this column is the key pressed. This process is repeated for all the rows.

Block diagram: Figure 7.40 shows the block diagram

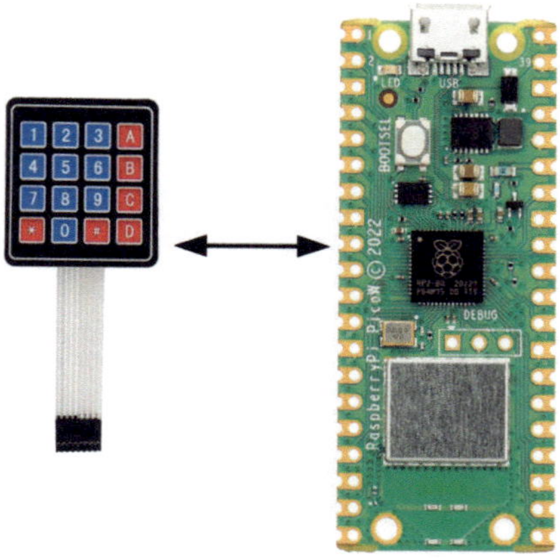

Figure 7.40: Block diagram.

Circuit diagram: The circuit diagram of the project is shown in Figure 7.41. The TFT display is connected to Pico as in the previous projects. The 4×4 keypad is connected to the following GPIO pins of the Raspberry Pi Pico W. The row pins are held high using the internal pull-up resistors to +3.3 V:

Keypad pin	**Raspberry Pi pin**
R1	P16
R2	GP17
R3	GP18
R4	GP19

C1	GP20
C2	GP21
C3	GP22
C4	GP26

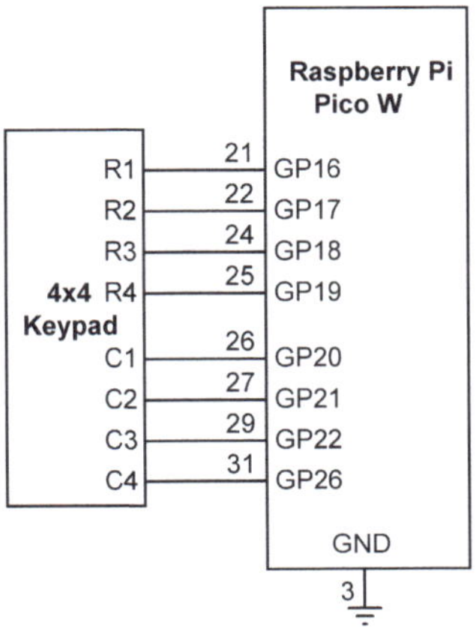

Figure 7.41: Circuit diagram.

Figure 7.42 shows the pin configuration of the 4×4 keypad used in the project.

Figure 7.42: Pin configuration of the 4×4 keypad.

Program listing: The operation of the keypad algorithm can be described by the following Program Description Language (PDL):

```
Configure all columns as outputs
Configure all rows as inputs
Set all columns to 1
DO for all columns
       Set a column to 0
       DO for all rows
              IF a row is 0 THEN
                     Return the key at this column and row position
              ENDIF
       ENDDO
ENDDO
```

Figure 7.43 shows the program listing (program: **keypad.py**). At the beginning of the program, the keypad keys are defined after importing the required modules to the program. The keypad row and columns connections are defined using lists **ROWS** and **COLS,** respectively. Columns are then configured as outputs and are set to 1. Similarly, the rows are configured as inputs. Function **Get_Key** reads the pressed key and returns it to the calling program. Two **for** loops are used in the function: the first loop selects the columns and sets them to 0 one after the other one. The second loop scans the rows and checks if a row is at 0. The main program calls the function and displays the pressed key on the screen. You should evaluate the keypad hardware and software by pressing various keys on the keypad and verifying that the correct key is displayed on the PC screen.

```
#---------------------------------------------------------------
#                    KEYPAD TEST PROGRAM
#                    ===================
#
# This program shows how the a keypad can be used to display
# the pressed keys
#
# Author: Dogan Ibrahim
# File  : keypad.py
# Date  : October 2022
#--------------------------------------------------------------
from machine import Pin
import utime

C = [0]*4
R = [0]*4

KEYPAD = [                                          # Keypad keys
    [1,2,3,"A"],
    [4,5,6,"B"],
    [7,8,9,"C"],
    ["*",0,"#","D"]]
```

```
ROWS = [16,17,18,19]                                # Row pins
COLS = [20,21,22,26]                                # Column pins

for i in range(4):                                  # Conf columns
   C[i] = Pin(COLS[i], Pin.OUT)
   C[i].value(1)

for j in range(4):                                  # Conf rows
   R[j] = Pin(ROWS[j], Pin.IN, Pin.PULL_UP)

#
# This function reads a key from the keypad
#
def Get_Key():
   while True:
      for j in range(4):
          C[j].value(0)                             # Set col j to 0
          for i in range(4):                        # For all rows
              if R[i].value() == 0:                 # Row is 0?
                    return (KEYPAD[i][j])           # Return key
                    while R[i].value() == 0:
                       pass
          C[j].value(1)                             # Col back to 1
          utime.sleep(0.05)                         # Wait 0.05s

while True:
   key = Get_Key()                                  # Get a key
   print(key)                                       # Display the key
   utime.sleep(0.5)
```

Figure 7.43: Program: **keypad.py**.

Pressing a button on the keypad displays its code on the screen. Figure 7.44 shows the display after pressing all the keys.

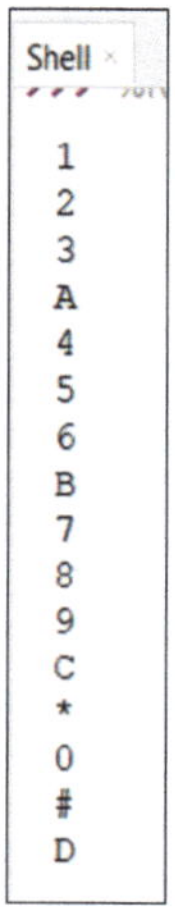

Figure 7.44: Pressing all the keys.

7.13 Project 9: Elementary multiplication – using 4×4 keypad and TFT

Description: This is an elementary multiplication exercise project. The program displays two random numbers between 1 and 100 and waits for the user to enter the result of multiplying these numbers. The result is checked, and feedback is given to the user such as **WELL DONE** or **WRONG...**The above process repeats after 5 seconds delay.

Aim: The aim of this project is for kids to exercise their multiplication skills.

Block diagram: Figure 7.45 shows the block diagram where a TFT display is added to the project.

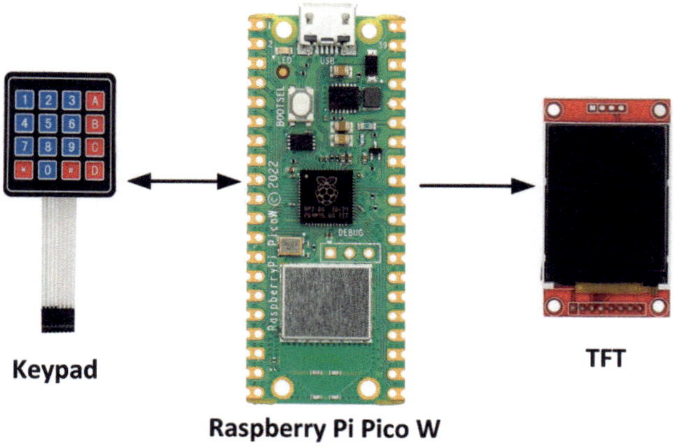

Figure 7.45: Block diagram of the project.

Circuit diagram: The circuit diagram of the project is shown in Figure 7.46. The TFT display is connected to the Pico as in the previous projects.

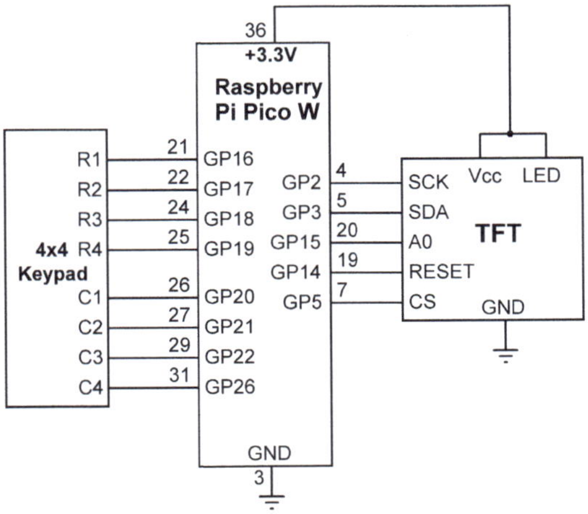

Figure 7.46: Circuit diagram of the project.

Program listing: Figure 7.47 shows the program listing (**TFTmultiply.py**). Key **D** is assumed to be the **ENTER** key and must be pressed after entering the answer. At the beginning of the program, all the required modules are imported. Function **Heading()** draws a red rectangle and displays **MULTIPLICATION**. Inside the main program, two random numbers are generated between 1 and 100 (**n1** and **n2**). The user is then expected to multiply these numbers by hand and enter the result. Key **D** must be entered after entering the result. The program checks the result and displays either **WELL DONE** or **WRONG...** The above process is repeated after 5 seconds when a new set of numbers are generated.

```
#-------------------------------------------------------------
#                   ELEMENTARY MULTIPLICATION
#                   =========================
#
# This is an elementary multiplication program. The program
# displays two random integer numbers between 1 and 100
# and expects the user to multiply these numbers correctly
#
# Author: Dogan Ibrahim
# File  : TFTmultiply.py
# Date  : October 2022
#-------------------------------------------------------------
import utime
from ST7735 import TFT
from sysfont import sysfont
from machine import SPI, Pin
import random

#
# Initialize the SPI
#
```

```
spi = SPI(0, baudrate=20000000, polarity=0, phase=0,
sck=Pin(2), mosi=Pin(3), miso=Pin(4))

tft = TFT(spi, 15, 14, 5)
tft.initg()
tft.rgb(True)

C = [0]*4
R = [0]*4

KEYPAD = [                                          # Keypad keys
    [1,2,3,"A"],
    [4,5,6,"B"],
    [7,8,9,"C"],
    ["*",0,"#","D"]]

ROWS = [16,17,18,19]                                # Row pins
COLS = [20,21,22,26]                                # Column pins

for i in range(4):                                  # Conf columns
   C[i] = Pin(COLS[i], Pin.OUT)
   C[i].value(1)

for j in range(4):                                  # Conf rows
   R[j] = Pin(ROWS[j], Pin.IN, Pin.PULL_UP)

def Heading():
    tft.fill(TFT.WHITE)                             # Clear screen
    tft.rect((5, 10), (110, 120), TFT.RED)
    tft.text((17, 30), "MULTIPLICATION", TFT.BLUE, sysfont,
       1.1, nowrap=True)

#
# This function reads a key from the keypad
#
def Get_Key():
   while True:
      for j in range(4):
          C[j].value(0)                             # Set col j to 0
          for i in range(4):                        # For all rows
             if R[i].value() == 0:                  # Row is 0?
                   return (KEYPAD[i][j])            # Return key
                   while R[i].value() == 0:
                      pass
          C[j].value(1)                             # Col back to 1
          utime.sleep(0.05)                         # Wait 0.05s
```

```
#
# Start of MAIN program
#
while True:
   Heading()
   n1 = random.randint(1, 100)                  # Random no
   n2 = random.randint(1, 100)                  # Random no
   tft.text((15, 60), "{:d} X {:d} = ".format(n1,n2), TFT.BLUE,
    sysfont, 1.1, nowrap=True)

   ans = 0
   key = " "
   r = 15
   while True:
       key = Get_Key()                          # Get a key
       if key == "D":
          break
       tft.text((r, 75), "{:d}".format(int(key)), TFT.MAROON,
       sysfont, 1.1, nowrap=True)
       r = r + 6
       ans = 10*ans + int(key)                  # Display the key
       utime.sleep(0.5)
       result = n1 * n2
   if result == ans:
       tft.text((10, 90), "WELL DONE",TFT.GREEN,sysfont,2,nowrap=True)
   else:
       tft.text((10, 90), "WRONG...",TFT.MAROON,sysfont,2,nowrap=True)
   utime.sleep(5)
```

Figure 7.47: Program: **TFTmultiply.py**.

Figure 7.48 shows an example display where the answer was correct.

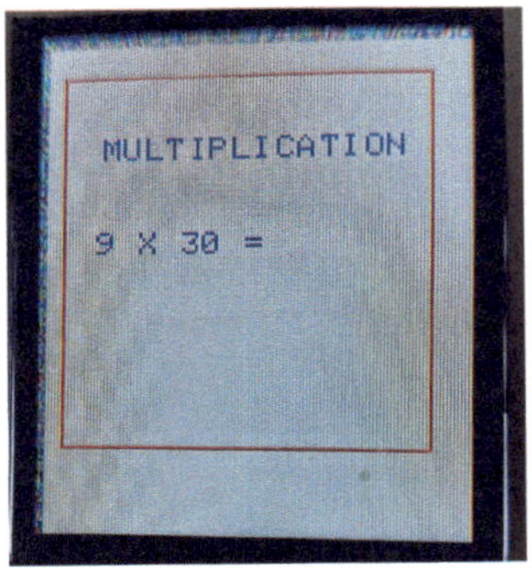

Figure 7.48: Example display.

7.14 Project 10: Calculator - using 4×4 keypad and TFT

Description: This is an integer calculator project. The user enters two integer numbers and the required operation (+ - * /). The result is calculated and displayed on the TFT display. The program repeats after 10 second's delay.

The block diagram and circuit diagram of the project are the same as in Figure 7.45 and Figure 7.46.

Program listing: Keypad key numbering for this project is shown below. Keys ABCs are replaced with +-*, key * with / and key D with ENTER:

```
1 2 3 +
4 5 6 -
7 8 9 *
/ 0 # ENTER
```

Figure 7.49 shows the program listing (Program: **TFTcalc**). Function **Get_Key()** reads a key from the keypad, and this is the same as in the previous project. The program prompts the user to enter two integer numbers. **while** loops are used to read the digits of the numbers entered by the user and these numbers are stored in variables **n1** and **n2**. These **while** loops are terminated when key **D** (ENTER) is pressed. The program then reads the required operation and stores it in variable **opkey**. The result is calculated depending on the value of **opkey**. For example, if addition is selected, then **n1** and **n2** are added and the result is displayed on the TFT.

```
#---------------------------------------------------------------
#                    CALCULATOR
#                    ==========
#
# This is an integer calculator program that can do the 4 basic
# operations of +-*/
#
# Author: Dogan Ibrahim
# File  : TFTcalc.py
# Date  : October 2022
#--------------------------------------------------------------
import utime
from ST7735 import TFT
from sysfont import sysfont
from machine import SPI, Pin
import random

#
# Initialize the SPI
#
spi = SPI(0, baudrate=20000000, polarity=0, phase=0,
```

```
sck=Pin(2), mosi=Pin(3), miso=Pin(4))

tft = TFT(spi, 15, 14, 5)
tft.initg()
tft.rgb(True)

C = [0]*4
R = [0]*4

KEYPAD = [                                  # Keypad keys
    [1,2,3,"+"],
    [4,5,6,"-"],
    [7,8,9,"*"],
    ["/",0,"#","D"]]

ROWS = [16,17,18,19]                        # Row pins
COLS = [20,21,22,26]                        # Column pins

for i in range(4):                          # Conf columns
   C[i] = Pin(COLS[i], Pin.OUT)
   C[i].value(1)

for j in range(4):                          # Conf rows
   R[j] = Pin(ROWS[j], Pin.IN, Pin.PULL_UP)

def Heading():
    tft.fill(TFT.WHITE)                     # Clear screen
    tft.rect((5, 10), (110, 120), TFT.RED)
    tft.text((26, 30), "CALCULATOR", TFT.BLUE, sysfont,
       1.1, nowrap=True)

#
# This function reads a key from the keypad
#
def Get_Key():
   while True:
      for j in range(4):
          C[j].value(0)                     # Set col j to 0
          for i in range(4):                # For all rows
              if R[i].value() == 0:         # Row is 0?
                    return (KEYPAD[i][j])   # Return key
                    while R[i].value() == 0:
                       pass
          C[j].value(1)                     # Col back to 1
          utime.sleep(0.05)                 # Wait 0.05s
#
```

```
# Start of MAIN program
#
while True:
   Heading()
#
# Get first number
#
   tft.text((10, 50), "First no: ", TFT.MAROON,sysfont, 1.1, nowrap=True)
   n1 = 0
   key = " "
   r = 70
   while True:
       key = Get_Key()                                  # Get a key
       if key == "D":
          break
       tft.text((r, 50), "{:d}".format(int(key)), TFT.MAROON,
       sysfont, 1.1, nowrap=True)
       r = r + 6
       n1 = 10*n1 + int(key)                            # Display the key
       utime.sleep(0.5)
#
# Get second number
#
   r = 70
   tft.text((10, 70), "Second no: ", TFT.MAROON,sysfont, 1.1, nowrap=True)
   n2 = 0
   key = " "
   r = 75
   while True:
       key = Get_Key()                                  # Get a key
       if key == "D":
          break
       tft.text((r, 70),"{:d}".format(int(key)),TFT.
MAROON,sysfont,1.1,nowrap=True)
       r = r + 6
       n2 = 10*n2 + int(key)                            # Display the key
       utime.sleep(0.5)

#
# Get operation
#
   tft.text((10, 90),"OP (+-*/): ",TFT.MAROON,sysfont,1.1,nowrap=True)
   key = " "
   while True:
       key = Get_Key()                                  # Get a key
       if key == "D":
```

```
            break
        tft.text((74, 90),"{:s}".format(key),TFT.MAROON,sysfont,1.1,nowrap=True)
        opkey = key
        utime.sleep(0.5)
#
# Display result
#
    if opkey == "+":
        result = n1 + n2
    elif opkey == "-":
        result = n1 - n2
    elif opkey == "*":
        result = n1 * n2
    elif opkey == "/":
        result = int(n1 / n2)
    tft.text((10, 110),"Result={:d}".format(result),TFT.
BLUE,sysfont,1.1,nowrap=True)

    utime.sleep(10)
```

Figure 7.49: Program: ***TFTcalc****.*

Figure 7.50 shows an example display.

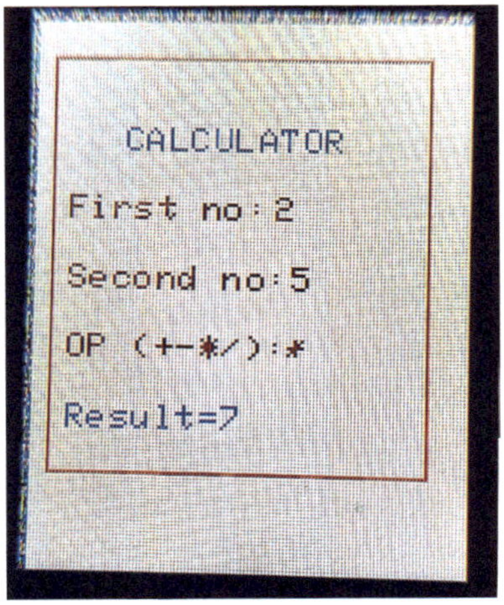

Figure 7.50: Example display.

7.15 Project 11: HiLo game - using 4×4 keypad and TFT

Description: This is the classical HiLo game. The computer generates a secret random number between 1 and 100 and the user tries to guess this number. The user is given hints when the entered guess is high or low.

The block diagram and circuit diagram of the project are the same as in Figure 7.45 and Figure 7.46.

Program listing: Figure 7.51 shows the program listing (Program: **HiLo.py**). The Keypad layout is the same as in Project 9 where **D** is the ENTER key. The program displays **TOO HIGH** if the difference between your guess and the secret number is higher than 50, displays **HIGH** if the guess is high but the difference is less than 50, displays **TOO LOW** if the difference between your guess and the secret number is lower than 50, and displays LOW if your guess is low but the difference is not lower than 50. The program also displays the number of attempts made when the secret number is found.

```
#------------------------------------------------------------------
#                    HiLo GAME
#                    ==========
#
# A random number is generated between 1 and 100 and the user
# attempts to guessthis number. Program gives hints as follows:
#
# TOO HIGH: The number guessed is very high (more than 50)
# HIGH: The number guessed is HIGH (less than 50)
# TOO LOW: The number guessed is too low (more than 50)
# LOW: The number guessed is low (less than 50)
#
# Author: Dogan Ibrahim
# File  : HiLo.py
# Date  : October 2022
#-----------------------------------------------------------------
import utime
from ST7735 import TFT
from sysfont import sysfont
from machine import SPI, Pin
import random


#
# Initialize the SPI
#
spi = SPI(0, baudrate=20000000, polarity=0, phase=0,
sck=Pin(2), mosi=Pin(3), miso=Pin(4))

tft = TFT(spi, 15, 14, 5)
tft.initg()
tft.rgb(True)

C = [0]*4
R = [0]*4
```

```
KEYPAD = [                                          # Keypad keys
    [1,2,3,"A"],
    [4,5,6,"B"],
    [7,8,9,"C"],
    ["*",0,"#","D"]]

ROWS = [16,17,18,19]                                # Row pins
COLS = [20,21,22,26]                                # Column pins

for i in range(4):                                  # Conf columns
   C[i] = Pin(COLS[i], Pin.OUT)
   C[i].value(1)

for j in range(4):                                  # Conf rows
   R[j] = Pin(ROWS[j], Pin.IN, Pin.PULL_UP)

def Heading():
    tft.fill(TFT.WHITE)                             # Clear screen
    tft.rect((5, 10), (110, 120), TFT.RED)
    tft.text((13, 30), "GUESS MY NUMBER", TFT.BLUE, sysfont,
       1.1, nowrap=True)

#
# This function reads a key from the keypad
#
def Get_Key():
   while True:
      for j in range(4):
          C[j].value(0)                             # Set col j to 0
          for i in range(4):                        # For all rows
              if R[i].value() == 0:                 # Row is 0?
                    return (KEYPAD[i][j])           # Return key
                    while R[i].value() == 0:
                       pass
          C[j].value(1)                             # Col back to 1
          utime.sleep(0.05)                         # Wait 0.05s


#
# Start of MAIN program
#
attempts = 0
n = random.randint(1, 100)

while True:
   Heading()
   tft.text((10, 50), "Your Guess: ", TFT.MAROON,sysfont, 1.1, nowrap=True)
```

```
    attempts = attempts + 1

    n1 = 0
    key = " "
    r = 80
    while True:
        key = Get_Key()                         # Get a key
        if key == "D":
           break
        tft.text((r, 50), "{:d}".format(int(key)), TFT.MAROON,
        sysfont, 1.1, nowrap=True)
        r = r + 6
        n1 = 10*n1 + int(key)                   # Display the key
        utime.sleep(0.5)

    if n1 == n:
       tft.text((10, 100),"WELL DONE",TFT.BLUE,sysfont,1.1,nowrap=True)
       tft.text((10, 120), "Attempts:{:d}".format(attempts),TFT.BLUE,
       sysfont,1.1,nowrap=True)
       attempts = 0
       n = random.randint(1, 100)
    elif n1 > n + 50:
       tft.text((10, 110),"TOO HIGH",TFT.BLUE,sysfont,1.1,nowrap=True)
    elif n1 > n:
       tft.text((10, 110),"HIGH",TFT.BLUE,sysfont,1.1,nowrap=True)
    elif n1 < n - 50:
       tft.text((10, 110),"TOO LOW",TFT.BLUE,sysfont,1.1,nowrap=True)
    elif n1 < n:
       tft.text((10, 110),"LOW",TFT.BLUE,sysfont,1.1,nowrap=True)
    utime.sleep(3)
```

Figure 7.51: Program: **HiLo.py**.

Figure 7.52 shows an example display.

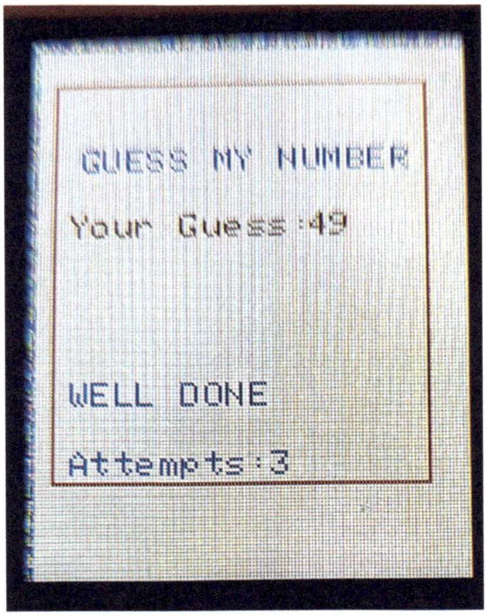

Figure 7.52: Example display.

Chapter 8 • I²C Bus Projects

8.1 Overview

The I²C bus (also written as: I²C) is commonly used in microcontroller-based projects. In this chapter, you shall be looking at the use of this bus on the Raspberry Pi Pico. The aim is to make the reader familiar with the I²C bus library functions and to show how they can be used in a real project. Before looking at the details of the project, it is worthwhile to look at the basic principles of the I²C bus.

8.2 The I²C Bus

As discussed briefly in Section 5.3, the I²C bus is one of the most used microcontroller communication protocols for communicating with external devices such as sensors and actuators. The I²C bus is a single master, multiple slave bus and it can operate at standard mode: 100 Kbit/s, full speed: 400 Kbit/s, fast mode: 1 Mbit/s, and high speed: 3.2 Mbit/s. The bus consists of two open-drain wires, pulled-up with resistors:

SDA: data line
SCL: clock line

Figure 8.1 shows the structure of an I²C bus with one master and three slaves. Notice that the correct operation of the bus requires pull-up resistors to the power supply. This is because the devices on the bus are in open-drain mode. Some slave devices already have internal pull-up resistors. It is important to check the devices used and if there are no pull-up resistors then external ones must be added when the device is on the bus.

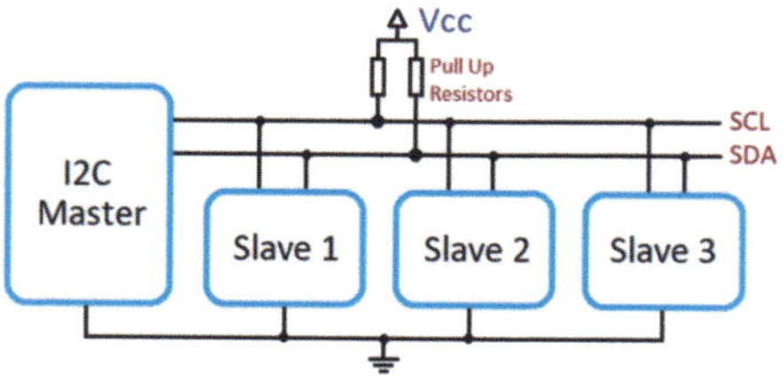

Figure 8.1: I²C bus elementary structure.

8.3 I²C pins of the Raspberry Pi Pico W

As discussed in Section 5.4, Raspberry Pi Pico and Pico W have two I²C pins, named I2C0 and I2C1, repeated here in Figure 8.2. As shown in the figure, the I²C pins are duplicated and shared with other pins. For example, GP0 (pin 1) is the I2C0 SDA pin and GP1 (pin 1) is the I2C0 SCL pin. Also, GP16 (pin 21) is the I2C0 SDA pin and GP17 (pin 22) is the I2C SCL pin.

Figure 8.2: Raspberry Pi Pico and Pico W I^2C pins.

The default I^2C pins are:

I2C0 SCL	GP9
I2C0 SDA	GP8
I2C1 SCL	GP7
I2C1 SDA	GP6

The master on the bus always initiates communication and usually requires data from a slave device (e.g., a sensor). Slave devices respond to the master and they cannot initiate communication on the bus. In addition to normal data transfer on the bus, an acknowledge bit (ACK) is used to signal the transmitter that a byte has been successfully received.

The advantages of using the I^2C bus are:

- Only 2 wires are needed.
- Up to 1008 slave devices are supported on the bus.
- Multi-master (more than one master) operation is supported.

The disadvantages of the I^2C bus are:

- The data rate is not very high.
- The bus timing is complex.

In the remaining parts of this chapter, you will be developing projects using the I^2C bus.

8.4 Project 1: I²C port expander

Description: A simple project is given in this section to show how the I²C functions can be used in a program. In this project, the I²C bus-compatible Port Expander chip (MCP23017) is used to give additional 16 I/O ports to the Raspberry Pi Pico W. This is useful in some applications where many I/O ports may be required. In this project, an LED is connected to MCP23017 port pin GPA0 (pin 21) and the LED is flashed ON and OFF every second so that the operation of the program can be verified. A 470-ohm current-limiting resistor is used in series with the LED.

The aim: The aim of this project is to show how the I²C bus can be used in Raspberry Pi Pico projects.

Block diagram: The block diagram of the project is shown in Figure 8.3.

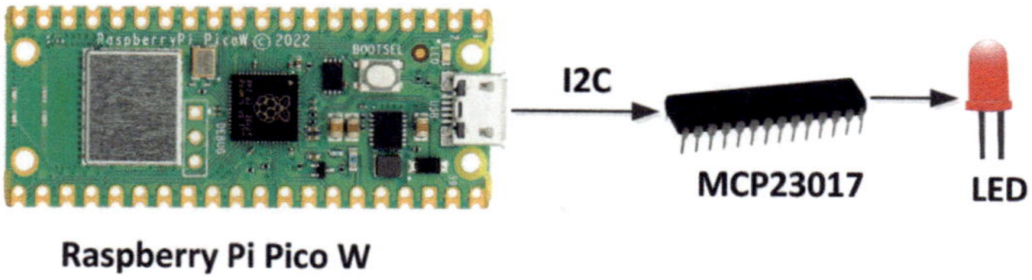

Figure 8.3: Block diagram of the project.

The MCP23017

The MCP23017 is a 28-pin chip with the following features. The pin configuration is shown in Figure 8.4:

- 16 bi-directional I/O ports
- Up to 1.7 MHz operation on the I²C bus
- Interrupt capability
- External reset input
- Low standby current
- +1.8 to +5.5 V operation
- 3 address pins so that up to 8 devices can be used on the I²C bus
- 28-pin DIL package

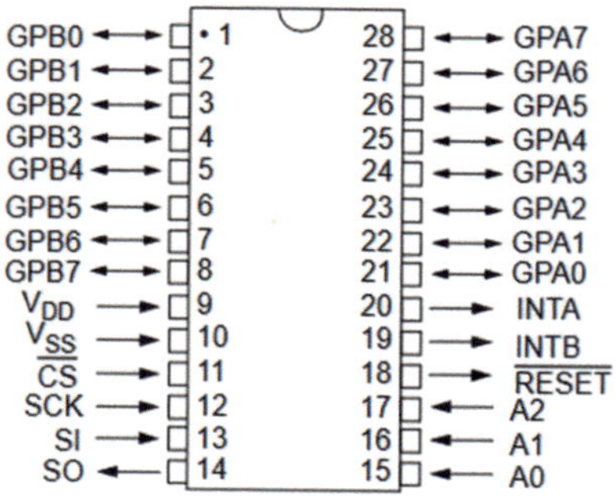

Figure 8.4: Pin configuration of the MCP23017.

The pin descriptions are given in Table 8.1.

Pin	Description
GPA0-GPA7	Port A pins
GPB0-GPB7	Port B pins
VDD	Power supply
VSS	Ground
SDA	I^2C data pin
SCL	I^2C clock pin
RESET	Reset pin
A0-A2	I^2C address pins

Table 8.1: MCP23017 pin descriptions.

The MCP23017 is addressed using pins A0 to A2. Table 8.2 shows the address selection. In this project, the address pins are connected to ground, thus the address of the chip is 0x20. The chip address is 7-bits wide with the low bit set or cleared depending on whether you wish to read data from the chip or write data to the chip respectively. Since in this project you will be writing to the MCP23017, the low bit should be 0, making the chip byte address (also called the **device opcode**) as 0x40.

A2	A1	A0	Address
0	0	0	0x40
0	0	1	0x21
0	1	0	0x22
0	1	1	0x23
1	0	0	0x24
1	0	1	0x25
1	1	0	0x26
1	1	1	0x27

Table 8.2: Address selection of the MCP23017.

The MCP23017 chip has 8 internal registers that can be configured for its operation. The device can either be operated in 16-bit mode or in two 8-bit modes by configuring bit IO-CON.BANK. On power-up, this bit is cleared which chooses the two 8-bit modes by default.

The I/O direction of the port pins is controlled with registers IODIRA (at address 0x00) and IODIRB (at address 0x01). Clearing a bit to 0 in these registers makes the corresponding port pin(s) output(s). Similarly, setting a bit to 1 in these registers makes the corresponding port pin(s) input(s). GPIOA and GPIOB register addresses are 0x12 and 0x13 respectively. This is shown in Figure 8.5.

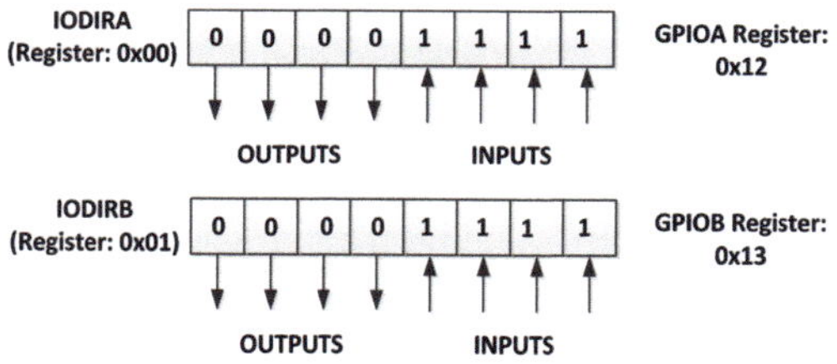

Figure 8.5: Configuring the I/O ports.

Figure 8.6 shows the circuit diagram of the project. Notice that the I²C pins of the port expander are connected to pins GP8 (I2C0 SDA) and GP9 (I2C0 SCL) of the Raspberry Pi Pico and are pulled up using 10 k-ohms resistors as required by the I²C specifications. The LED is connected to port pin GPA0 of the MCP23017 (pin 21). The address select bits of the MCP23017 are all connected to ground.

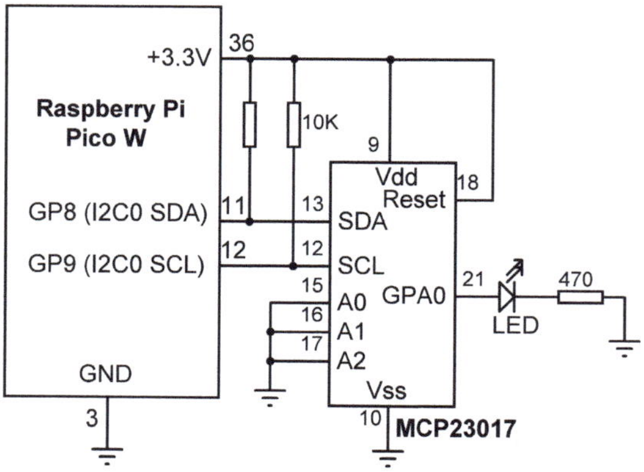

Figure 8.6: Circuit diagram of the project.

More information on the MCP23017 chip can be obtained from the datasheet:

http://docs-europe.electrocomponents.com/webdocs/137e/0900766b8137eed4.pdf

Program listing: Figure 8.7 shows the program listing (Program: **MCP23017**). Raspberry Pi Pico supports the following I²C functions:

i2c.scan()	scan for slave I²C devices
i2c.writeto()	write to I²C bus
i2c.readfrom()	read from I²C bus
i2c.writeto_mem()	write to memory of slave device
i2c.readfrom_mem()	read from memory of slave device

Some examples of I^2C operations are:

`print("I2C address=", i2c.scan())`	print the I2C slave addresses on the bus
`i2c.writeto(0x20, b'56')`	write 56 to I2C address 0x20
`i2c.writeto(0x20, bytearray(buff))`	write buffer to I2C address 0x20
`i2c.readfrom(0x20, 3)`	read 3 bytes from I2C address 0x20
`i2c.writeto_mem(0x20, 0x10, b'\x35')`	write 0x35 to memory address 0x10 of the slave whose I2C address is 0x20
`i2c.readfrom_mem(0x20, 0x10, 3)`	read 3 bytes starting from register address 0x10 of slave I2C device whose address is 0x20

At the beginning of the program, the I^2C module is imported to the program and the I^2C interface is defined by specifying the SDA and SCL pin connections to the Pico. Function **i2c.scan()** is called to display the I^2C slave devices on the bus and the following was displayed:

i2c address = [32] which corresponds to hexadecimal 0x20

Then the MCP23017 I^2C device address, register **GPIOA** address, and the I/O direction **IODIRA** address are defined. List **conf** stores the **IODIRA** register address and 0 so that MCP23017 PORTA is set to output mode. List **buff1** stores the **GPIOA** register address and what the output should be set to (1 to turn ON the LED). Similarly, **buff2** stores the **GPIOA** register address and what the output should be set to (0 to turn OFF the LED). Inside the main program loop, the LED is flashed every second.

```
#-----------------------------------------------------------
#               I2C PORT EXPANDER
#               =================
#
# In this project the MCP23017 port expander chip is used.
# An LED is connected to port pin GP0 of the port expander
# and this LED is flashed every second
#
# Author: Dogan Ibrahim
# File  : MCP23017.py
# Date  : October 2022
#-------------------------------------------------------------
from machine import Pin,I2C
import utime

i2c = machine.I2C(0, scl=Pin(9), sda=Pin(8), freq=100000)
```

```
print("i2c address=",i2c.scan())

Device_Address = 0x20                    # MCP23017 I2C address
MCP_GPIOA_REG = 0x12                     # MCP23017 GPIOA address
MCP_IODIRA_REG = 0                       # MCP23017 IODIRA Address

conf = [MCP_IODIRA_REG, 0]               # Configure as output
buff1 = (MCP_GPIOA_REG, 0)               # Set GPIOA to 0
buff0 = [MCP_GPIOA_REG, 1]               # Set GPIOA to 1

i2c.writeto(Device_Address, bytearray(conf))

while True:
    i2c.writeto(Device_Address, bytearray(buff1))
    utime.sleep(1)
    i2c.writeto(Device_Address, bytearray(buff0))
    utime.sleep(1)
```

Figure 8.7: Program: ***MCP23017****.*

The program can be modified so that the memory functions of I²C are used. The modified program listing (Program: **MCP23017-2**) is shown in Figure 8.8.

```
#-------------------------------------------------------------
#              I2C PORT EXPANDER
#              =================
#
# In this project the MCP23017 port expander chip is used.
# An LED is connected to port pin GP0 of the port expander
# and this LED is flashed every second
#
# This version of the program uses the i2c memory functions
#
# Author: Dogan Ibrahim
# File  : MCP23017-2.py
# Date  : October 2022
#---------------------------------------------------------------
from machine import Pin,I2C
import utime

i2c = machine.I2C(0, scl=Pin(9), sda=Pin(8), freq=100000)
print("i2c address=",i2c.scan())

Device_Address = 0x20                    # MCP23017 I2C address
MCP_GPIOA_REG = 0x12                     # MCP23017 GPIOA address
MCP_IODIRA_REG = 0                       # MCP23017 IODIRA Address
```

```
i2c.writeto_mem(Device_Address, MCP_IODIRA_REG, b'0')

while True:
    i2c.writeto_mem(Device_Address, MCP_GPIOA_REG, b'1')
    utime.sleep(1)
    i2c.writeto_mem(Device_Address, MCP_GPIOA_REG, b'0')
    utime.sleep(1)
```

Figure 8.8: Modified program: **MCP23017-2**.

8.5 Project 2: TMP102 temperature sensor with LCD

Description: In this project, the I2C -compatible TMP102 temperature sensor chip is used. The ambient temperature is read every 3 seconds and is displayed on the I^2C display.

The aim: The aim of this project is to show how the temperature sensor chip TMP102 can be used in a program.

The TMP102

The TMP102 is an I^2C -compatible highly accurate temperature sensor chip with a built-in thermostat, having the following basic features:

Supply voltage: 1.4 V to 3.6 V
Supply current: 10 µA
Accuracy: ±0.5 ^oC
Resolution: 12 bits (0.0625 ^oC)
Operating range: –40 ^oC to +125 ^oC

TMP102 is a 6-pin chip as shown in Figure 8.9. the pin descriptions are:

Pin	Name	Description
1	SCL	I^2C clock line
2	GND	power supply ground
3	ALERT	Over temperature alert. Open-drain output. requires a pull-up resistor
4	ADD0	Address select
5	V+	power supply
6	SDA	I^2C data line

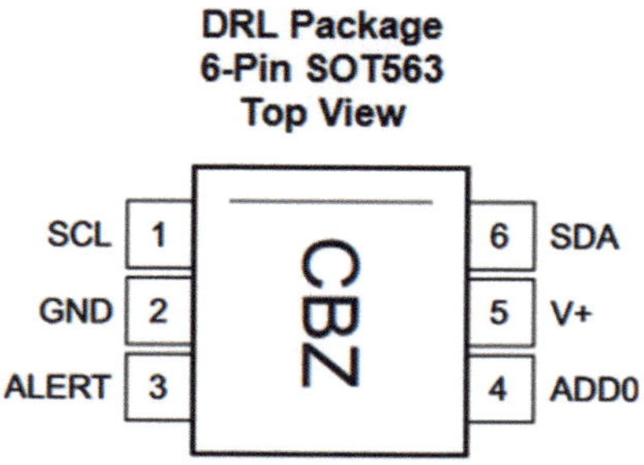

Figure 8.9: TMP102 pin layout.

TMP102 has the following operational modes:

- **Continuous conversion**: by default, an internal ADC converts the temperature into digital format with the default conversion rate of 4 Hz, with a conversion time of 26 ms. The conversion rate can be selected using bits **CR1** and **CR0** of the configuration register as: 0.25 Hz, 1 Hz, 4 Hz (default), and 8 Hz. In this project, the default 4 Hz is used.
- **Extended mode**: Bit EM of the configuration register selects normal mode (EM = 0), or extended mode (EM = 1). In normal mode (default mode), the converted data is 12 bits. Extended mode is used if the temperature is above 128 ºC and the converted data is 13 bits. In this project, the normal mode is used.
- **Shutdown mode**: This mode is used to save power where the current consumption is reduced to less than 0.5 µA. The shutdown mode is entered when the configuration register bit SD = 1. The default mode is normal operation (SD = 0).
- **One-shot conversion**: Setting configuration register bit OS to 1 selects the one-shot mode which is a single conversion mode. The default mode is continuous conversion (OS = 0).
- **Thermostat mode**: This mode indicates whether to operate in comparator mode (TM = 0) or in interrupt mode (TM = 1). The default is the comparator mode. In comparator mode, the Alert pin is activated when the temperature equals or exceeds the value in the THIGH register and remains active until the temperature drops below TLOW. In interrupt mode, the Alert pin is activated when the temperature exceeds THIGH or goes below TLOW registers. The Alert pin is cleared when the host controller reads the temperature register.

A **Pointer Register** selects various registers in the chip as shown in Table 8.3. The upper 6 bits of this register are 0s.

P1	P0	REGISTER SELECTED
0	0	Temperature register (read only)
0	1	Configuration register
1	0	TLOW register
1	1	THIGH register

Table 8.3: Pointer register bits.

Table 8.4 shows the temperature register bits in normal mode (EM = 0).

BYTE 1:

D7	D6	D5	D4	D3	D2	D1	D0
T11	T10	T9	T8	T7	T7	T5	T4

BYTE 2:

D7	D6	D5	D4	D3	D2	D1	D0
T3	T2	T1	T0	0	0	0	0

Table 8.4: Temperature register bits.

Table 8.5 shows the configuration register bits. The power-up default bit configuration is shown in the Table.

BYTE 1:

D7	D6	D5	D4	D3	D2	D1	D0
OS	**R1**	**R0**	**F1**	**F0**	**POL**	**TM**	**SD**
0	1	1	0	0	0	0	0

BYTE 2:

D7	D6	D5	D4	D3	D2	D1	D0
CR1	**CR0**	**AL**	**EM**	**0**	**0**	**0**	**0**
1	0	1	0	0	0	0	0

Table 8.5: Configuration register bits.

The **Polarity** bit (POL) allows the user to adjust the polarity of the Alert pin output. If set to 0 (default), the Alert pin becomes active low. When set to 1, the Alert pin becomes active high.

The default device address is 0x48. TMP102 is available as a module (breakout module) as shown in Figure 8.10. The temperature register address is 0x00 and this should be sent after sending the device address. This is then followed with a read command where 2 bytes are read from the TMP102. These 2 bytes contain the temperature data.

The temperature read sequence is as follows:

- Master sends the device address 0x48 with the R/W set to 0.
- Device responds with ACK.
- Master sends the temperature register address 0x00.
- Device responds with ACK.
- Master re-sends device address 0x48 with the R/W bit set to 1.
- Master reads the upper byte of temperature data.
- Device sends ACK.
- Master reads the lower byte of temperature data.
- Device sends ACK.
- Master sends stop condition on the bus.

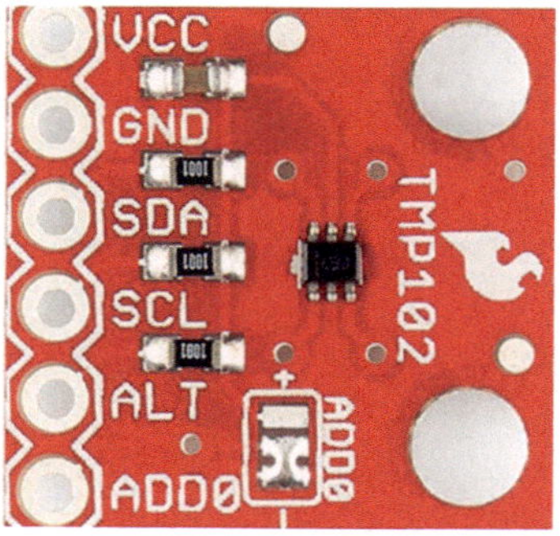

Figure 8.10: TMP102 as a module.

Block diagram: Figure 8.11 shows the block diagram of the project.

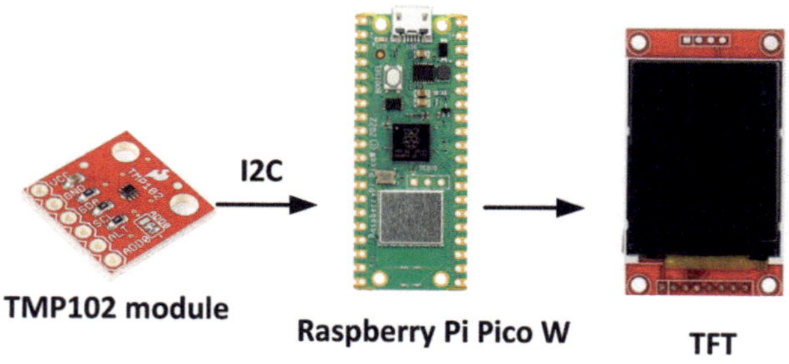

Figure 8.11: Block diagram of the project.

Circuit diagram: The circuit diagram of the project is shown in Figure 8.12. On-chip pull-up resistors are available on the TMP102 I^2C bus lines and therefore there is no need to use external pull-up resistors.

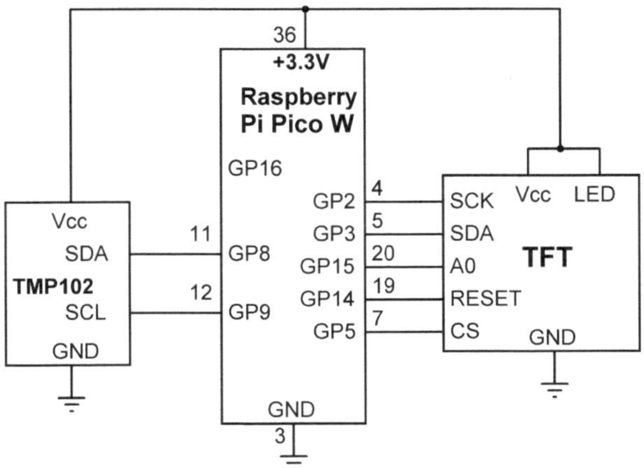

Figure 8.12: Circuit diagram of the project.

Program listing: Figure 8.13 shows the program listing (Program: **TMP102.py**). At the beginning of the program, the I^2C address of TMP102 and the Pointer register addresses are defined. The Pointer register is set to 0 to select the temperature register.

The program runs inside a **while** loop and calls function **Read()**. This function reads the temperature from TMP102, converts it into positive (or negative) degrees Celsius and then returns the reading to the main program. The two bytes read are combined to form the 12-bit temperature data in variable **temp**. If the temperature is negative, then it is in 2's complement form and its complement is taken and 1 is added to find the true negative value. By multiplying **temp** with the LSB, you find the temperature in degrees Centigrade. The temperature is displayed on the I^2C LCD as a floating-point number.

Table 8.6 shows the data output format of the temperature. Let us look at two examples:

Example 1: Measured value = 0011 00100000 = 0x320 = 800 decimal

This is a positive temperature, so the temperature is 800 × 0.0625 = +50 ^{o}C.

Example 2: Measured value = 1110 01110000 = 0xE70

This is negative temperature, complement is 0001 10001111, adding 1 gives 0001 10010000 = 400 decimal. The temperature is 400 × 0.0625 = 25 or, –25 ^{o}C

Temperature	Digital Output (Binary)	Digital Output (Hex)
128	011111111111	7FF
100	011001000000	640
50	001100100000	320
0.25	000000000100	004
-0.25	111111111100	FFC
-25	111001110000	E70
-55	110010010000	C90

Table 8.6: The data output for some temperature readings.

```
#------------------------------------------------------------
#              TMP102 TEMPERATURE SENSOR
#              =========================
#
# In this project a TMP102 type I2C compatible temperature
# sensor chip is connected to Raspberry Pi Pico. The temperature
# readings are displayed on the I2C LCD.
#
# Author: Dogan Ibrahim
# File  : TMP102.py
# Date  : October 2022
#--------------------------------------------------------------
import utime
from ST7735 import TFT
from sysfont import sysfont
from machine import SPI, Pin, I2C

#
# Initialize the SPI
#
spi = SPI(0, baudrate=20000000, polarity=0, phase=0,
sck=Pin(2), mosi=Pin(3), miso=Pin(4))

tft = TFT(spi, 15, 14, 5)
tft.initg()
tft.rgb(True)

#
# Initialize I2C
#
i2c = machine.I2C(0, scl=Pin(9), sda=Pin(8), freq=100000)
print("i2c address=",i2c.scan())
```

```
Device_Address = 0x48                          # TMP102 I2C address
PointerReg = 0                                 # TMP102 register

#
# This function reads the temperature, extracts degrees Celius
# and returns the temperature to the main calling program
#
def Read():
    data = [0, 0]
    LSB = 0.0625
#    data = i2c.readfrom_mem(Device_Address, PointerReg, 2)
    temp = (data[0] << 4) | (data[1] >> 4)
    if temp > 0x7FF:
        temp = (~temp) & 0xFF
        temp = temp + 1
        temperature = -temp * LSB
    else:
        temperature = temp * LSB
    return(temperature)

#
# Main program reads and displays the temperature every 3 seconds
#
while True:
    Temperature = Read()
    tft.fill(TFT.WHITE)                                # Clear screen
    tft.rect((5, 10), (100, 120), TFT.RED)
    tft.text((17, 30),"TEMPERATURE",TFT.BLUE,sysfont,1.1,nowrap=True)
    tft.text((15, 60), "Temp:{:.2f}C".format(Temperature), TFT.BLUE,
    sysfont, 1.1, nowrap=True)
    utime.sleep(3)
```

Figure 8.13: Program: **TMP102.py**.

Chapter 9 • OLED Displays

9.1 Overview

In microcontroller-based systems, you usually want to interact with the system for example to enter a parameter, to change the value of a parameter, or to display the output of a measured variable. Data is usually entered into a system using a switch, a small keypad, or a fully blown keyboard. Data is usually displayed using an indicator such as one or more LEDs, 7-segment displays, TFT, OLED, or simple LCD-type displays. LCDs have the advantages that they are cheap and can display alphanumeric as well as graphical data. Some LCDs have 40 or more character lengths with the capability to display data in several lines. Some other LCDs can be used to display graphical images (Graphical LCDs, or simply GLCDs), such as animation. Some displays are in single or in multi-colour, while some others incorporate back lighting so that they can be viewed in dimly lit conditions. Other display devices such as TFT and OLEDs have the advantages that they can display colour graphical data. Additionally, OLEDs are cheap, and they do not require background lighting since each pixel on the display is capable of emitting light.

The OLED display used in this chapter is the 0.91-inch I^2C -compatible display (Figure 9.1). This is a small low-cost 128×32-pixel display. This display has the following basic features:

- Driver: SSD1306
- Interface: I^2C
- Display colour: White
- Resolution: 128×32
- Viewing angle: >160°
- Operating voltage: +1.65 to +3.3 V
- Dimension: 36 × 12.5 (mm)

Figure 9.1: 0.91-inch OLED display.

The XY coordinates of the used OLED display are shown in Figure 9.2.

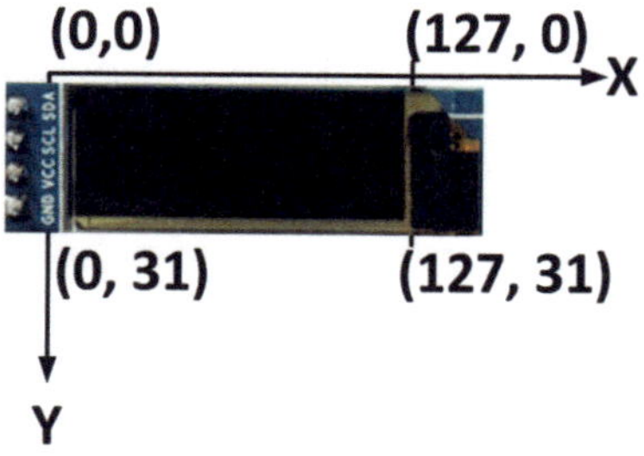

Figure 9.2: OLED display coordinates.

9.2 Installing the SSD1306 driver software

The SSD1306 driver library must be loaded to your Raspberry Pi Pico W before the library functions can be used. The library file is called **ssd1306.py** and is available on the Internet. This file is also available on the book's web site. Copy the file to your Raspberry Pi Pico with the exact name **ssd1306.py**.

9.3 Hardware interface

The OLED display is connected to Raspberry Pi Pico using the following pins (see Figure 9.3):

OLED Display	**Raspberry Pi Pico W**
SCL	GP9 (I2C0 SCL)
SDA	GP8 (I2C0 SDA)
GND	GND
VCC	+3.3V

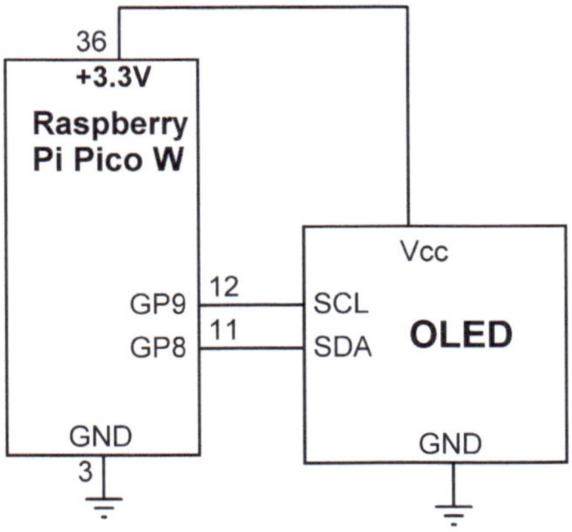

Figure 9.3: OLED – Pico interface.

You are now ready to use the library functions of the OLED display in your projects.

9.4 Displaying text on OLED

The program (**OLEDText.py**) shown in Figure 9.4 displays a 2-line text starting from column 5 of the display. Notice that at the beginning of the program library module ssd1306 is imported. Then the I^2C SDA and SCL connections of the OLED display are configured. The program displays the I^2C address of the display. Although this address information is not used in the program, it shows that the OLED is recognized by the Pico as an I^2C device. statement **oled.text()** displays text starting from the given pixel coordinates. Notice that the statement **oled.show()** must be included to refresh the display with new data.

```
#---------------------------------------------------------------
#                              OLED DISPLAY
#                              ============
#
# This is an OLED display test program. The following texts
# are displayed on the OLED:
#
#   Micropython (at 5,5)
#   Example     (at 5,25)
#
# Author: Dogan Ibrahim
# File  : OLEDText.py
# Date  : October 2022
#-----------------------------------------------------------------
from machine import Pin, I2C
from ssd1306 import SSD1306_I2C
import machine

i2c = I2C(0, scl=Pin(9),sda=Pin(8),freq=100000)
oled = SSD1306_I2C(128,32,i2c)         # Rows and columns

#
# Display device address (not used)
#
print("I2C Address: " + hex(i2c.scan()[0]).upper())

oled.fill(0)
oled.text("MicroPython",5,5)           # "MicroPython" at pixel (5,10)
oled.text("Example",5,25)              # "Example" at pixel (5,25)
oled.show()                            # Display
print("End")
```

Figure 9.4: Program: ***OLEDText.py****.*

Figure 9.5 shows the display when the above program is run.

Figure 9.5: The display.

When displaying text, it is easier to work with the display lines and columns. With the OLED display using the default character set, we can have up to 15 characters across and 3 characters down the display. i.e., **when displaying text using the 0.91-inch OLED, you can have 3 lines and 15 columns**. You can modify the code given above by creating a function called **display_text** so that lines and columns are entered for displaying text, instead of the raw pixel numbers as was the case with the above example:

```
        def display_text(text, line, column):
                oled.text(text, 8*column, 8*line)
```

The display function **fill(0)** sets all pixels to black and can be used to clear the screen if the background is black. Similarly, **fill(1)** sets all pixels to white and can be used to clear the screen if the background is white.

With the **oled.text** function you can also specify the text colour as the third argument of the function, where 0=black and 1=white.

You can now clear the screen and display text at line 1 and line 3 as shown in Figure 9.6 (Program: **OLEDText2.py**).

```
#-----------------------------------------------------------
#                              OLED DISPLAY
#                              ============
#
# This is an OLED display test program. The following texts
# are displayed on the OLED:
#
#   Micropython (at line 1, column 1)
#   Example     (at line 3, column 4)
#
# Author: Dogan Ibrahim
# File  : OLEDText2.py
# Date  : October 2022
#-------------------------------------------------------------
from machine import Pin, I2C
from ssd1306 import SSD1306_I2C
import machine

i2c = I2C(0, scl=Pin(9),sda=Pin(8),freq=100000)
oled = SSD1306_I2C(128,32,i2c)      # Rows and columns

def display_text(text, line, column):
      oled.text(text, 8*column, 8*line)

oled.fill(0)                         # Clear display
display_text("MicroPython",1,1)      # "MicroPython" at line 1 col 1
display_text("Program",3, 4)         # "Program" at line 3, col 4
oled.show()                          # Display
```

Figure 9.6: Program: ***OLEDText2.py****.*

In the above example, text **MicroPython** is displayed at line 1, starting from column 1, text **Program** is displayed at line 3, starting from column 4 (Figure 9.7).

Figure 9.7: The display.

9.5 Displaying common shapes

It is important to be able to draw various shapes using a graphical LCD such as the OLED. Some of the drawing functions are (note that colour is 1 for white, and 0 for black):

rect(xtop, ytop, width, height, color): This function draws an empty rectangle where **xtop** and **ytop** are the top left corners of the rectangle in pixels.

fill_rect(xtop, ytop, width, height, color): This function draws a filled rectangle where **xtop** and **ytop** are the top left corners of the rectangle in pixels.

line(xstrt, ystrt, xend, yend, color): This function draws a line starting from pixel position **(xstrt, ystrt)** to **(xend, yend)**.

hline(xstrt, ystrt, length, color): This function draws a horizontal line starting from pixel position **(xstrt, ystrt)** and having specified **length**

vline(xstrt, ystrt, length, color): This function draws a vertical line starting from pixel position **(xstrt, ystrt)** and having specified **length**

As an example, let us draw an empty rectangle, a line, a horizontal line, and a vertical line. The program for this example is shown in Figure 9.8 (Program: **OLEDshapes.py**). the following shapes are drawn:

```
oled.rect(2, 2, 120, 30, 1)          # Rectangle
oled.line(5, 5, 30, 25, 1)           # Line
oled.hline(20, 10, 50, 1)            # Horizontal line
oled.vline(50, 15, 12, 1)            # Vertical line
```

```
#------------------------------------------------------------
#              OLED SHAPES
#              ============
#
# This program displays various shapes on the OLED
```

```
#
# Author: Dogan Ibrahim
# File  : OLEDshapes.py
# Date  : October 2022
#-------------------------------------------------------------
from machine import Pin, I2C
from ssd1306 import SSD1306_I2C
import machine

i2c = I2C(0, scl=Pin(9),sda=Pin(8),freq=100000)
oled = SSD1306_I2C(128,32,i2c)          # Rows and columns

oled.fill(0)                            # Clear display
oled.rect(2, 2, 120, 30, 1)             # Rectangle
oled.line(5, 5, 30, 25, 1)              # Line
oled.hline(20, 10, 50, 1)               # Horizontal line
oled.vline(50, 15, 12, 1)               # Vertical line
oled.show()                             # Display
```

Figure 9.8: Program: **OLEDshapes.py**.

Figure 9.9 shows the display when the above program is run.

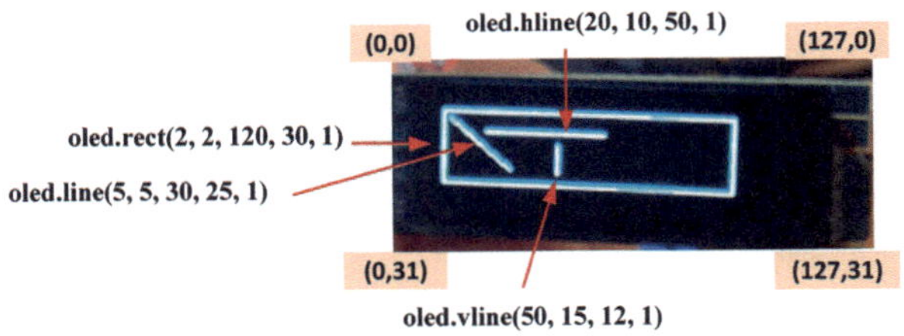

Figure 9.9: The display.

9.6 Other useful functions

Some other useful OLED functions are:

pixel(x, y, colour): Set pixel at **(x, y)** to specified colour (1=white, 0=black)

invert(1): This function inverts the colors so that black becomes white and white becomes black

invert(0): Return to the original colors

poweroff(): Power off the display (pixels remain in memory)

power(on): Power on the display. Pixels are redrawn

contrast(0): Dim the display

contrast(1): Display bright

rotate(True): This function rotates the display 180 degrees

rotate(False): Rotate the display 0 degrees

scroll(horiz, vert): This function scrolls the pixels **horiz** pixels to the right and **vert** pixels down.

Example - Circle

Write a function to draw a circle using the pixel function. The (x, y) coordinates of the center point, the radius, and the colour (black=0 or white=1) of the circle are to be passed as arguments. Set the (x, y) coordinates to 64, 30-pixel positions respectively, and the radius to 25 pixels.

Solution

In this example you will be using the following two formulas to draw the circle:

xi+1 = xi + r * cos(angle)
yi+1 = yi + r * sin(angle)

Where the angle takes values from 0 to 359, and r is the center pixel position. Figure 9.10 shows the required program (program: **OLEDcircle**). When this program is run the circle center at (64, 16) and radius 15 pixels is drawn as shown in Figure 9.11.

```
#===============================================================
#                        DRAW A CIRCLE USING PIXELS
#                        ==========================
#
# This program uses function circle to draw a circle with the
# given center coordinates and the radius in pixels
#
# Author: Dogan Ibrahim
# File  : OLEDcircle.py
# Date  : October, 2022
#===============================================================
from machine import Pin, I2C
from ssd1306 import SSD1306_I2C
import machine
import math

i2c = I2C(0, scl=Pin(9),sda=Pin(8),freq=100000)
oled = SSD1306_I2C(128,32,i2c)      # Rows and columns
```

```
#
# This function draws a circle using pixels
#
def circle(x, y, r, colour):
    for i in range(360):
        angle = i
        xi = r*math.cos(math.radians(angle))
        yi=r*math.sin(math.radians(angle))
        oled.pixel(int(x+xi),int(y+yi),colour)

#
# Circle radius at (64, 10), radius 15 pixels
#
oled.fill(0)                            # Clear display
circle(64, 16, 15, 1)                   # Circle spev
oled.show()                             # Draw
```

Figure 9.10: Program: **OLEDcircle**.

Figure 9.11: Displaying the circle.

Example - Square

Write a function to draw a square. The top left (x, y) coordinates, the length (or height) of the square, and its colour are to be passed as arguments to the function. Set the top left coordinates to (15, 15) and the length to 40 pixels.

Solution

The required program is shown in Figure 9.12 (program: **OLEDsquare**). Library function **rect** is used with the length and height set to the same value so that a square is drawn. When this program is run the square shown in Figure 9.13 is drawn.

```
#==============================================================
#                       DRAW A SQUARE
#                       =============
#
# This program draws a square where the top left x and y
# coordinates and the length of the required square are passed
# as arguments to the square function
#
# Author: Dogan Ibrahim
# File  : OLEDsquare.py
```

```
# Date  : October, 2022
#===============================================================
from machine import Pin, I2C
import ssd1306
import machine
import math

i2c = machine.SoftI2C(scl=Pin(9),sda=Pin(8),freq=100000)       #Init i2c
oled=ssd1306.SSD1306_I2C(128,32,i2c)                            # Rows and columns

#
# This function draws a square
#
def square(xtop, ytop, length, colour):
    oled.rect(xtop, ytop, length, length, colour)

oled.fill(0)                            # Clear display
square(10, 10, 20, 1)                   # Square
oled.show()                             # Draw
```

Figure 9.12: Program: ***OLEDsquare.***

Figure 9.13: Displaying the square.

Example – Scrolling shape

Write a program to scroll a square shape across the display horizontally by 5 pixels at every 0.5 seconds. Set the top left coordinates to (10, 10) and the length to 20 pixels.

Solution

The required program is shown in Figure 9.14 (program: **ScrollShape**). OLED function scroll is used to scroll the shape horizontally.

```
#===============================================================
#                      SCROLL A SHAPE
#                      ==============
#
# This program draws a square where the top left x and y
# coordinates and the length of the required square are passed
# as arguments to the square function. This square shape is then
# scrolled right across the display every 0.5 second
#
# Author: Dogan Ibrahim
```

```
# File  : ScrollShape.py
# Date  : October, 2022
#================================================================
from machine import Pin, I2C
import ssd1306
import machine
import utime

i2c = machine.SoftI2C(scl=Pin(9),sda=Pin(8),freq=100000)
oled=ssd1306.SSD1306_I2C(128,32,i2c)           # Rows and columns

#
# This function draws a square
#
def square(xtop, ytop, length, colour):
    oled.rect(xtop, ytop, length, length, colour)

oled.fill(0)                                   # Clear display
square(10, 10, 20, 1)                          # Square
oled.show()

#
# Scroll the shape across the display horizontally every 0.5s
#
for i in range(0, 90, 5):
    oled.scroll(5, 0)
    oled.show()
    utime.sleep(0.5)
```

Figure 9.14: Program: **ScrollShape**.

9.7 Project 1: Seconds counter

Description: This is a seconds counter project. A pushbutton switch is connected to one of the I/O ports. When the switch is pressed the program starts counting up every second and displays the count on OLED.

Aim: The aim of this project is to show how data can be updated on the OLED.

Circuit diagram: As shown in Figure 9.15, a pushbutton switch is connected to port pin GP15 (pin 20). The state of the pin is at logic 1 and goes to logic 0 when the switch is pressed.

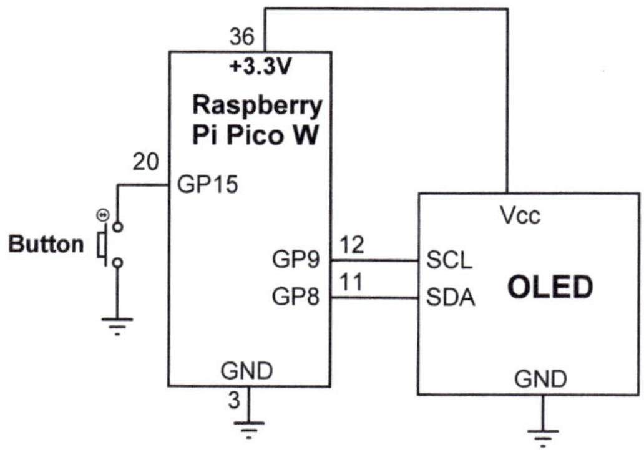

Figure 9.15: Circuit diagram of the project.

Program listing: Figure 9.16 shows the program listing (program: **SecondsCounter**). At the beginning of the program the libraries used in the program are imported, function **display_text** is defined, and the OLED is configured. GP15 is then configured as an input to which the button is connected. The program then displays the heading **Press Button...** and waits until the button is pressed. When the button is pressed the remainder of the program runs in a **while** loop where variable **Count** is incremented every second and displayed on the display.

```
#----------------------------------------------------------
#               SECONDS COUNTER
#               ===============
#
# In this program a pushbutton switch is connected to IO23.
# When the button is pressed the program starts counting up
# every second and displays the count on the OLED
#
# Author: Dogan Ibrahim
# File  : SecondsCounter.py
# Date  : October, 2022
#-----------------------------------------------------------
from machine import Pin, I2C
import ssd1306
import machine
import utime


#
# Function to display text at specified line and column
#
def display_text(text, line, column):
       oled.text(text, 8*column, 8*line)

i2c = machine.SoftI2C(scl=Pin(9),sda=Pin(8),freq=100000)
```

```
oled=ssd1306.SSD1306_I2C(128,32,i2c)            # Rows and columns

button=Pin(15,Pin.IN, pull=Pin.PULL_UP)         # Button at GP15
count = 0                                       # Count=0

oled.fill(0)                                    # Clear display
display_text("Press Button...", 2, 1)           # Display heading
oled.show()

while(button.value() == 1):                     # Wait for button
    pass

while True:
    oled.fill(0)                                # Clear display
    display_text("Count: ", 2, 1)               # Display "Count:"
    count = count + 1                           # Increment Count
    display_text(str(count), 2, 7)              # Display count
    oled.show()
    utime.sleep(1)                              # Wait 1 second
```

Figure 9.16: Program: **SecondsCounter***.*

Figure 9.17 shows the display before and after the button is pressed.

Figure 9.17: Display before and after the button is pressed.

9.8 Project 2: Drawing bitmaps

Description: There are many applications where you may want to draw images, e.g., bitmaps. Examples are given in this section to show how bitmaps can be drawn on OLED.

Simple shapes

As an example, let us draw a number of heart shapes across the OLED display. The bitmap of a heart shape is shown below. This shape for example can be extracted by using the **Paint** program:

```
ICON = [
[ 0, 0, 0, 0, 0, 0, 0, 0, 0],
[ 0, 1, 1, 0, 0, 0, 1, 1, 0],
[ 1, 1, 1, 1, 0, 1, 1, 1, 1],
[ 1, 1, 1, 1, 1, 1, 1, 1, 1],
[ 1, 1, 1, 1, 1, 1, 1, 1, 1],
```

```
[ 0, 1, 1, 1, 1, 1, 1, 1, 0],
[ 0, 0, 1, 1, 1, 1, 1, 0, 0],
[ 0, 0, 0, 1, 1, 1, 0, 0, 0],
[ 0, 0, 0, 0, 1, 0, 0, 0, 0],
]
```

Program listing: Figure 9.18 shows the program (program: **OLEDheart**) that draws 6 heart shapes across the display as shown in Figure 9.19.

```
#------------------------------------------------------------
#                       DISPLAY A BITMAP IMAGE
#                       ======================
#
# In this program heart images are displayed across the OLED
#
# Author: Dogan Ibrahim
# File  : OLEDheart.py
# Date  : October, 2022
#--------------------------------------------------------------
from machine import Pin, I2C, ADC
import ssd1306
import machine

i2c = machine.SoftI2C(scl=Pin(9),sda=Pin(8),freq=100000)
oled=ssd1306.SSD1306_I2C(128,32,i2c)    # Rows and columns

#
# Heart shape bitmap
#
ICON = [
[ 0, 0, 0, 0, 0, 0, 0, 0, 0],
[ 0, 1, 1, 0, 0, 0, 1, 1, 0],
[ 1, 1, 1, 1, 0, 1, 1, 1, 1],
[ 1, 1, 1, 1, 1, 1, 1, 1, 1],
[ 1, 1, 1, 1, 1, 1, 1, 1, 1],
[ 0, 1, 1, 1, 1, 1, 1, 1, 0],
[ 0, 0, 1, 1, 1, 1, 1, 0, 0],
[ 0, 0, 0, 1, 1, 1, 0, 0, 0],
[ 0, 0, 0, 0, 1, 0, 0, 0, 0],
]

#
# Clear and display 6 heart shapes across the display
#
oled.fill(0)
```

```
for i in range(0, 120, 20):
 for y, row in enumerate(ICON):
    for x, c in enumerate(row):
        oled.pixel(x+i, y+20, c)

oled.show()
```

Figure 9.18: Program listing.

Figure 9.19: Displayed heart shapes by the program.

Creating and displaying a bitmap

In this section, you will create a 128×32 bitmap image and then show it on the display. The image you will create is the handwriting of the letter A. The steps are given below:

Creating the image

You can use image creation programs such as Photoshop, GIMP, Microsoft Windows Paint and so on. In this example, you will be using the Microsoft program Paint to create your image.

- Start the **Paint** program.
- Click **Home → Resize** and click **Pixels**. Set the **Horizontal** to 128 pixels and the **Vertical** to 32 pixels as shown in Figure 9.20. Click **OK**.

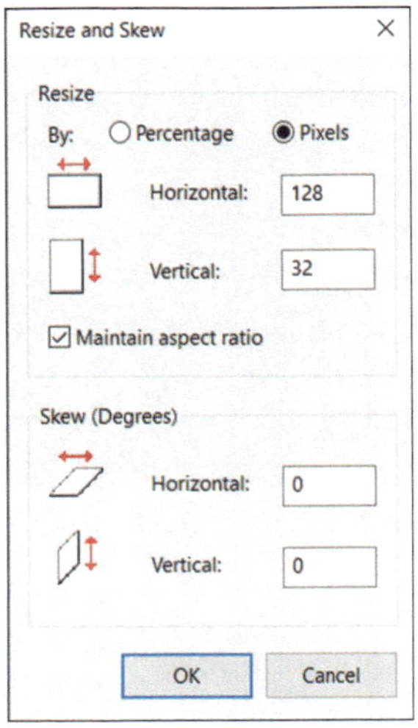

Figure 9.20: Set the pixels to 128×32.

- Click **View** and then click **Zoom In** to enlarge the grid.

- Draw the shape you want using the mouse. In this project, letter **A** is drawn by hand as shown in Figure 9.21.

Figure 9.21: Draw the required shape.

- Save the file, e.g., as **LetterA.png** in a folder on your PC. Exit from the Paint program.

- Now you have to convert the image to the correct format. The steps are:

- Start program **GIMP** and open your file **LetterA.png** (Figure 9.22).

Figure 9.22: Open the file in GIMP.

- Click **Image** → **Mode** → **Indexed** and select Use black and white (1-bit palette (Figure 9.23). Click **Convert**.

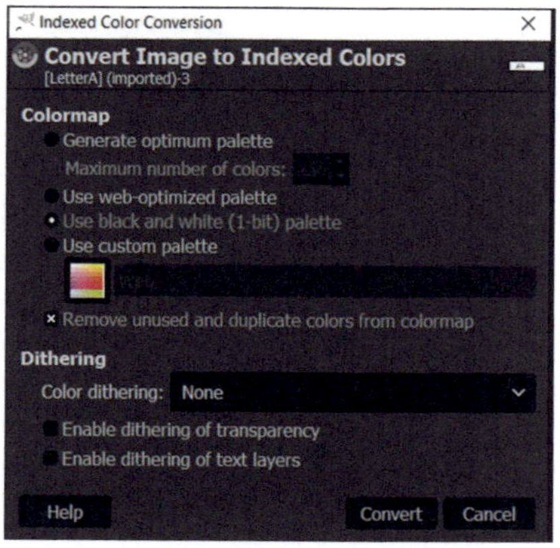

Figure 9.23: Select a 1-bit palette with black and white.

- Click **File** → **Export As** and then change the file extension to **pbm** at the top left corner of the screen. Click **Export**.

- Select **Raw** as the **Data formatting** and click **Export**. A new file called **LetterA.pbm** will be at your selected folder. Exit **GIMP**.

- Now, copy your image file from the PC to your Pico. Open Thonny, click **This Computer** at the left and select file **LetterA.pbm** (Figure 9.24).

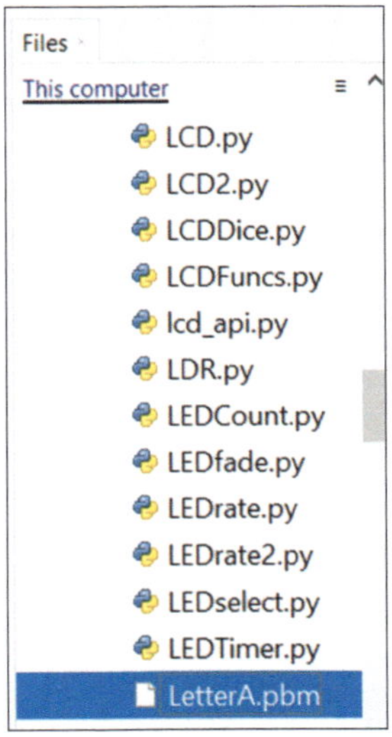

Figure 9.24: Select file **LetterA.pbm**.

- Right-click on the selected file and click **Upload to/**. Check Thonny and click **File → Open** and select **Raspberry Pi Pico** device to make sure that file **LetterA.pbm** has been copied to your Pico.

Program listing: Now that you have the bitmap image file on your development board, the next step is to write the program to display this image on the OLED. Figure 9.25 shows the program listing (program: **OLEDbitmap**). At the beginning of the program, the OLED interface is defined. The program then opens and reads the image file **LetterA.pbm**. Notice that the OLED **framebuffer** is used in this example. The SSD1306 OLED display has a display RAM which is same size as the display pixel size and there is one to one relationship between the memory bits and the pixels on the display. The framebuffer object has the same size as the display RAM and this object allows the programmer to load the framebuffer memory image before actually writing the data to the memory image to the display RAM. This way, you can for example load the framebuffer with the image you want to display and then load the display from the framebuffer. OLED function **blit()** is used to copy data from the framebuffer memory to the display as a block copy. Normally, a white image will be displayed on the OLED (see Figure 9.26 with the black background. You can invert the image so that the image is black and the background while as shown in Figure 9.26).

```
#-------------------------------------------------------------
#                       DISPLAY A BITMAP FILE
#                       ======================
#
# In this program a bitmap file is displayed on the OLED. The
# file was created using the Windows Paint program and then
# formatted using the GIMP
#
# Author: Dogan Ibrahim
# File  : OLEDbitmap.py
# Date  : October, 2022
#---------------------------------------------------------------
from machine import Pin, I2C
import ssd1306
import machine
import framebuf

i2c = machine.SoftI2C(scl=Pin(9),sda=Pin(8),freq=100000)
oled=ssd1306.SSD1306_I2C(128,32,i2c)    # Rows and columns

#
# Clear screen. open and read the file and display the image
#
oled.fill(0)
with open("LetterA.pbm", "rb") as bitmap:
    bitmap.readline()
    bitmap.readline()
    bitmap.readline()
    bitdata = bytearray(bitmap.read())

frbuf = framebuf.FrameBuffer(bitdata, 128, 32, framebuf.MONO_HLSB)
#oled.invert(1)
oled.blit(frbuf, 0, 0)
oled.show()
```

Figure 9.25: Program: **OLEDBitmap.py***.*

Figure 9.26 shows the display after the program is run.

Figure 9.26: The display before and after inverting it.

9.9 Project 3: DS18B20 OLED-based digital thermometer

Description: In this project, you will design a thermometer using the DS18B20 digital temperature sensor chip and display the ambient temperature on the OLED every second.

Aim: The aim of this project is to show how the DS18B20 temperature sensor chip can be used in a project and how the temperature can be displayed every second on the OLED.

DS18B20: This is a digital thermometer chip that provides 9-bit to 12-bit Celsius temperature measurements and additionally it includes an alarm function with programmable trigger points. The chip communicates over a 1-Wire bus. It has a 64-bit serial code, which allows multiple DS18B20s to function on the same 1-Wire bus.

The basic features of the DS18B20 chip are:

- Measures temperature from –55 ºC to +125 ºC.
- It has ±0.5 ºC accuracy from –10 ºC to +85 ºC.
- Programmable resolution from 9 bits to 12-bits.
- No external components required.
- Has a unique 64-bit serial code stored on-chip ROM.

This is a 3-pin module with the following pin definitions:

- VCC: power supply pin (+3.3 V)
- DQ: input/output pin
- GND: power supply ground pin

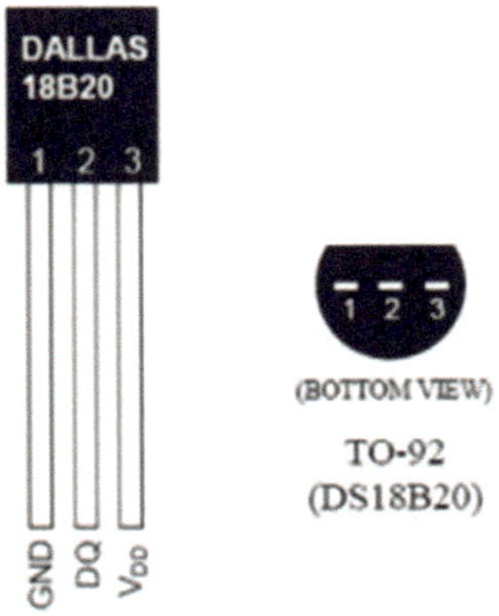

Figure 9.27: DS18B20 module.

Block diagram: Figure 9.28 shows the block diagram. The pointer needle indicates the temperature on a paper scale.

Figure 9.28: Block diagram of the project.

Circuit diagram: Figure 9.29 shows the circuit diagram of the project. The DQ pin of the DS18B20 module is connected to pin GP15 of Pico.

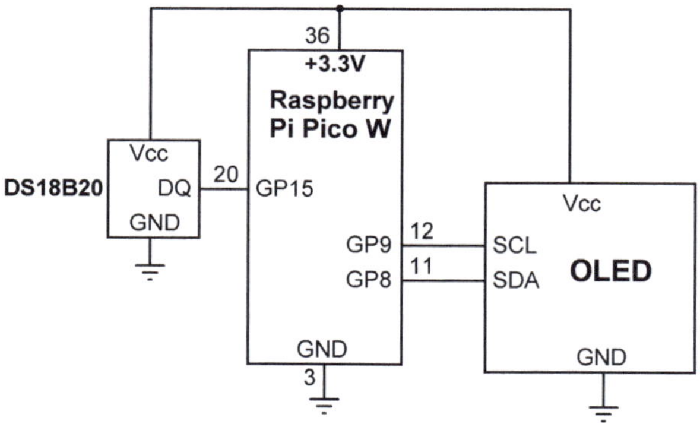

Figure 9.29: Circuit diagram of the project.

Program listing: Figure 9.30 shows the program listing (program: **OLEDtemp**). At the beginning of the program, the OLED configuration and connection to the DS18B20 chip are defined. Then, an endless loop is formed using a **while** statement. Inside this loop, the temperature is read and displayed on the OLED every second.

```
#-----------------------------------------------------------
#                       DIGITAL THERMOMETER
#                       ===================
#
# In this program a DS18B20 digital temperature sensor is
# connected to the development board. The ambient temperature
# is measured every scond and displayed on the OLED
#
# Author: Dogan Ibrahim
# File  : OLEDtemp.py
# Date  : October, 2022
#------------------------------------------------------------
from machine import Pin, I2C
import ssd1306
```

```
import machine
import utime
import onewire, ds18x20

#
# Function to display text at specified line and column
#
def display_text(text, line, column):
       oled.text(text, 8*column, 8*line)

i2c = machine.SoftI2C(scl=Pin(9),sda=Pin(8),freq=100000)
oled=ssd1306.SSD1306_I2C(128,32,i2c)            # Rows and columns

Temp = machine.Pin(15)                          # DS18B20 at GP15
ds = ds18x20.DS18X20(onewire.OneWire(Temp))

#
# Read the temperature and display on OLED
#
while True:
    roms=ds.scan()                              # Scan chip
    ds.convert_temp()                           # Convert temperature
    utime.sleep_ms(750)                         # Wait for conversion

    for rom in roms:
        temperature = ds.read_temp(rom)         # Read temperature

    oled.fill(0)                                # Clear display
    display_text("Temp(C)= ", 3, 1)             # Display "Count:"
    display_text(str(temperature)[:5], 3, 9)    # Display count
    oled.show()
    utime.sleep(1)                              # Wait 1 second
```

Figure 9.30: Program listing.

A typical temperature display is shown in Figure 9.31, where the measured temperature was 23.62 ºC at the time of measurement.

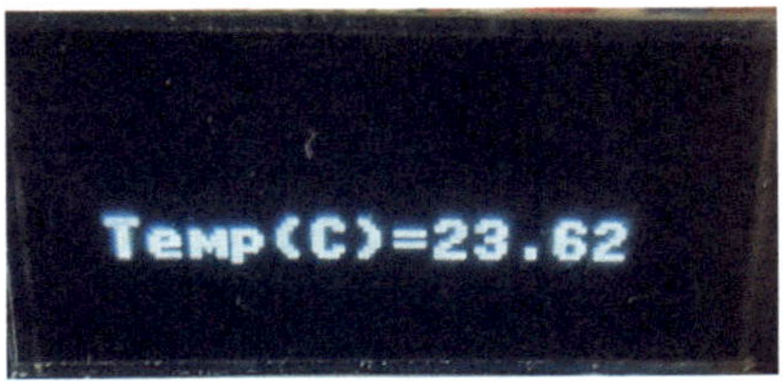

Figure 9.31: Typical display of the temperature.

9.10 Project 4: Heart rate (pulse) measurement

Description: It is important to know the heart rate, especially during an exercise or activity. The Pulse Sensor is a small analog device that can be used to measure the heart rate easily. The device includes a simple optical sensor with amplification and noise-cancellation circuitry so that reliable readings can be obtained.

A pulse sensor is a small device (Figure 9.32) with a 3-pin connector, and it has the following pins:

-	GND
S	signal output
Middle	+3.3 V

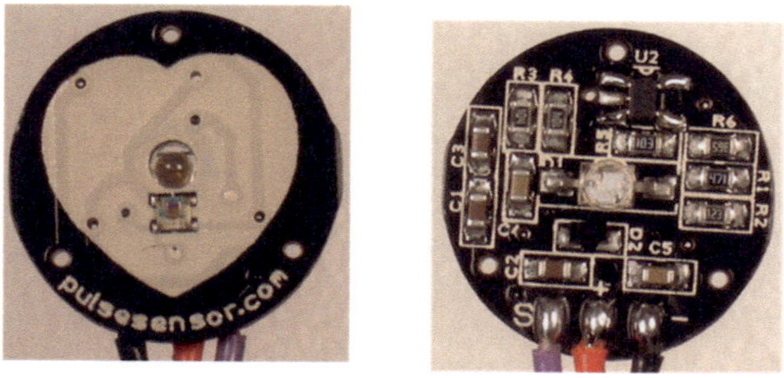

Figure 9.32: Pulse sensor module.

The sensor can be held between fingers. Alternatively, you can purchase additional parts such as ear clips or vinyl stickers so that the sensor can be attached to the earlobe. On the front of the sensor is a heart logo and this is the side that contacts the skin. There is a small hole where the LED shines through from the back. On the side, a small light sensor pad is placed. The remaining parts are mounted at the rear of the sensor.

In this project, you will detect the heart rate and then display it on the OLED.

> **Note:** it is recommended to be safe and not to connect the Pulse Sensor to your body while your development kit is being powered from the main AC line through an adaptor.

Block diagram: Figure 9.33 shows the block diagram of the project.

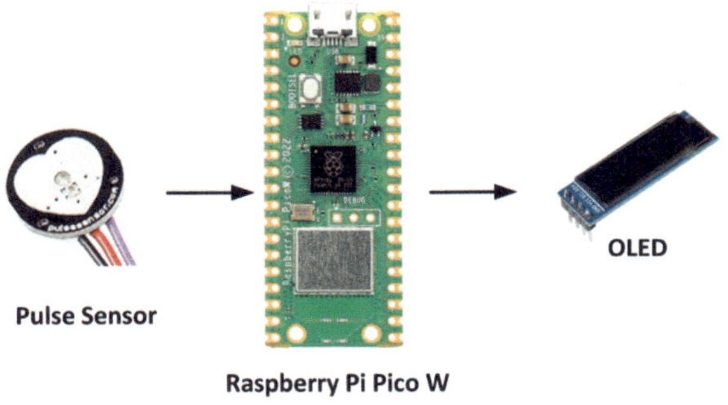

Figure 9.33: Block diagram of the project.

Circuit diagram: The circuit diagram is shown in Figure 9.34. The output (pin **S**) of the sensor is connected to analog input GP26 (ADCON0) of Pico. Additionally, GND and VCC are connected to GND and +3.3 V pins respectively.

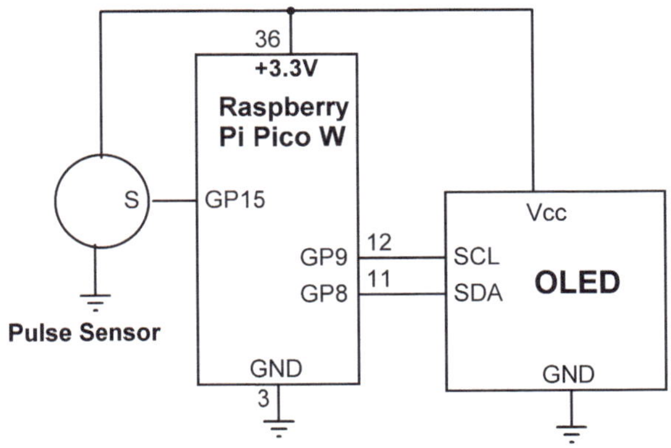

Figure 9.34: Circuit diagram of the project.

Program listing: The program listing is shown in Figure 9.35 (program: **OLEDpulse**). At the beginning of the project, the OLED and the pulse sensor interfaces are defined. Then a message is displayed on the OLED and the user is prompted to place his/her finger on the pulse sensor. The program carries out a calibration and calculates the maximum and minimum values read by the ADC. These values are stored in variables **maxv** and **minv** respectively. The remaining of the code runs in an endless loop setup using a **while** statement. Inside this loop, the output of the pulse sensor is read, and the BPM is incremented if the signal exceeds a threshold, where this threshold is used to calculate the BPM. The program then displays the values read from the ADC while the display is scrolled left by one pixel so that a graph of the data can be drawn in real-time. The BPM is then calculated and displayed at the top part of the display. Notice that this part of the display is scrolled left as well. The above process is repeated until stopped by the user.

In this project, the value of the BPM was calibrated using a professional pulse sensor device (e.g., a Pulse Oximeter). You may need to re-calibrate with your own setup. Notice that the results obtained here may not be accurate.

```
#------------------------------------------------------------
#                HEART RATE (PULSE) MEASUREMENT
#                ==============================
#
# In this project a Pukse Sensor module is connected to analogue
# input IO32 of the development board. The program detects
# the heart rate pulses and displays a graph on the OLED. Also,
# the Beats Per Minute (BPM) is calculated and displaed
#
# Author: Dogan Ibrahim
# File  : OLEDpulse.py
# Date  : October, 2022
#--------------------------------------------------------------
from machine import Pin, I2C, ADC
import ssd1306
import machine
import utime
import math

i2c = machine.SoftI2C(scl=Pin(9),sda=Pin(8),freq=100000)
oled=ssd1306.SSD1306_I2C(128,32,i2c)   # Rows and columns

AnalogIn = ADC(0)                      # ADC channel 0

minmax = [0]*200
def display_text(text, line, column):
       oled.text(text, 8*column, 8*line)

display_text("Heart Rate Mtr", 1, 1)   # Display heading
display_text("Finger on...", 2, 1)
oled.show()
utime.sleep(3)                         # Give time to user
display_text("Wait...", 3, 1)
oled.show()

#
# Calculate maximum and minimum values received by ADC
#
for i in range(200):
    minmax[i] = AnalogIn.read_u16()
    utime.sleep_ms(10)

minv = min(minmax)                     # Minimum
maxv = max(minmax)                     # MAximum
diff = maxv-minv                       # Difference
```

```
oled.fill(0)                              # Clear display
oled.show()
lasty=30
bpm = 0
cnt = 0

#
# Read the heart rate and display on the OLED
#
while True:
    adcValue = AnalogIn.read_u16()        # Read temp
    if adcValue >= maxv - 30000:          # Over the threshold
       bpm = bpm + 1
    y = 30 - int((30*(adcValue - minv))/diff)
    oled.line(125,lasty,126,y,1)
    oled.scroll(-1,0)                     # Scroll left
    lasty=y
    oled.show()
    utime.sleep_ms(20)                    # Wait 10 ms
    cnt = cnt + 1
    if cnt == 60:
       cnt = 0
       oled.fill_rect(0,0,30,12,0)        # Clear area
       oled.text("%d BPM" % bpm, 30, 0)           # Display BPM
       bpm = 0
```

Figure 9.35: Program listing.

Figure 9.36 shows the display when the program is started. The user is expected to place his/her finger on the sensor when the text **Finger on...** is displayed. The figure also shows the graph of the heart rate and display of the BPM.

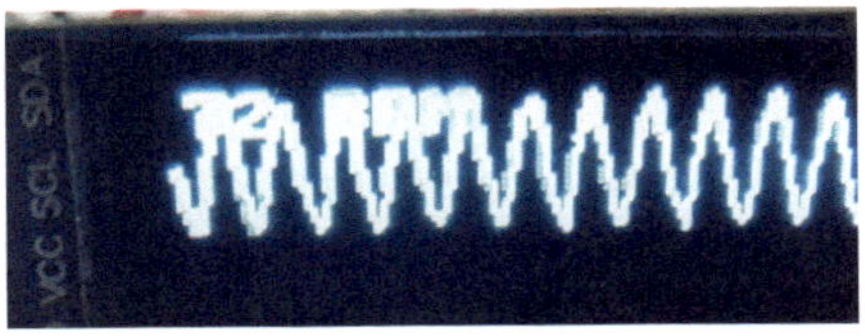

Figure 9.36: Example display.

Chapter 10 • Using Bluetooth with the Raspberry Pi Pico W

10.1 Overview

Bluetooth is one of the popular means of exchanging data wirelessly over short distances. Nowadays, many electronic devices such as smartphones, laptops, iPads, games, gadgets, portable health monitoring devices, and so on are all equipped with Bluetooth modules. Bluetooth is used by many people to share pictures and music files using their smartphones.

Bluetooth is a paired communication protocol where both devices must enable their Bluetooth links and use then use the same key to connect to each other. When the connection is established, the data can be sent both ways. There is no need to worry about line-of-sight between the devices since the communication is based on radio waves.

Sometimes the pairing between devices may fail. You should pay attention to the following points for successful pairing between the devices:

- Make sure Bluetooth is turned on, on both devices.
- Make your device discoverable. On some devices, you may have to click a button to make Bluetooth discoverable.
- Male sure the two devices are close to each other.
- Make sure the devices to be paired are compatible with each other, e.g., their versions are compatible.
- Enter the same pairing code on both devices when asked.

10.2 Raspberry Pi Pico W Bluetooth interface

The Wi-Fi module on the Raspberry Pi Pico W supports Bluetooth communications. Unfortunately, at the time of writing this book, the Bluetooth firmware was not ready and as a result, Bluetooth support was not available. Hopefully, future releases will support Bluetooth communications.

Because Bluetooth support is not (yet) available on the Raspberry Pi Pico W, you may consider using an external Bluetooth module to enable the Pico to communicate with other Bluetooth devices. Perhaps the cheapest and the easiest option is to use a serial Bluetooth module such as the HC-06. In the next sections, you will develop projects and learn how to connect an HC-06-type low-cost Bluetooth module to your Raspberry Pi Pico W.

10.3 Project 1: Controlling three LEDs from a smartphone using Bluetooth

Description: In this project, you will be sending commands over the Bluetooth link from a smartphone to control three LEDs connected to the Raspberry Pi Pico W (you could easily replace the LEDs with other components such as buzzers, relays, etc. so that electrical devices can be controlled remotely). In this project, the three LEDs are named R, G. and B. Valid commands are:

R1	Turn R LED ON
R0	Turn R LED OFF
G1	Turn G LED ON
G0	Turn G LED OFF
B1	Turn B LED ON
B0	Turn B LED OFF

Aim: The aim of this project is to show how to use a low-cost serial Bluetooth module with the Raspberry Pi Pico W to control electrical devices remotely (in this project 3 LEDs).

The HC-06 Bluetooth module

The HC-06 is a low-cost popular 4-pin serially-controlled module with the following pins (see Figure 10.1):

Figure 10.1: The HC-06 Bluetooth module.

The HC-06 has the following basic specifications:

- +3.3 V to +6 V operation
- 30 mA unpaired current (10 mA matched current)
- Built-in antenna
- Band: 2.40 GHz – 2.48 GHz
- Power level: +6 dBm
- Default communication: 9600 baud, 8 data bits, no parity, 1 stop bit
- Signal coverage 30 feet
- Safety feature: Authentication and encryption
- Modulation mode: Gaussian frequency-shift keying

Block Diagram: Figure 10.2 shows the block diagram of the project.

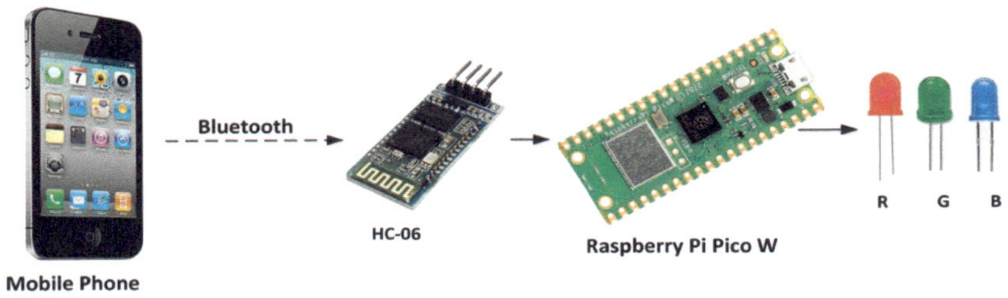

Figure 10.2: Block diagram of the project.

Circuit Diagram: Figure 10.3 shows the project circuit diagram. RXD and TXD pins of the Bluetooth module are connected to UART 0 pins GP0 (TX) and GP1 (RX) of Raspberry Pi Pico respectively. The LEDs are connected to GP14 (LED R), GP15 (LED G), and GP16 (LED B) through 470-ohm current-limiting resistors.

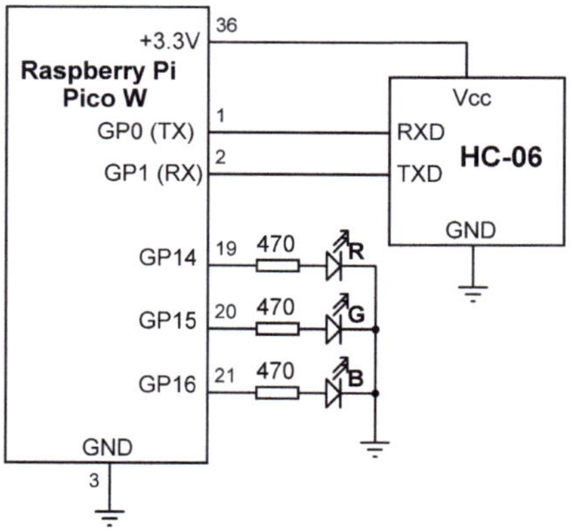

Figure 10.3: Circuit diagram of the project.

Program listing: The program listing of the project is shown in Figure 10.4 (program: **BlueLED**). At the beginning of the program, the hardware UART interface is set to baud rate 9600 which is the default speed of HC-06 and the LEDs are configured as outputs and are turned OFF. The remaining of the program runs in an endless loop. Inside this loop, data (commands) are received from the Bluetooth device using function call **readline**. The data read is stored in the list called **buf**. The program then controls the LEDs based on the received commands.

```
#-----------------------------------------------------------
#              BLUETOOTH COMMUNICATION
#              =======================
#
# In this project a HC-06 type serial Bluetooth module and
# and 3 LEDs are connected to the Raspberry Pi Pico.The LEDs
# are controlled by sending commands from a Bluetooth
# compatible smart phone. Confirmation is sent to smart phone
#
# Author: Dogan Ibrahim
# File  : BlueLED.py
# Date  : October 2022
#-------------------------------------------------------------
from machine import Pin, UART
import utime

uart = UART(0, baudrate=9600,rx=Pin(1),tx=Pin(0))

R = Pin(14, Pin.OUT)                   # R LED
```

```
G = Pin(15, Pin.OUT)                      # G LED
B = Pin(16, Pin.OUT)                      # Y LED

#
# Turn OFF LEDs
#
R.value(0)
G.value(0)
B.value(0)

#
# Main program loop. Receive a command and control the LED
#
while True:
     buf = uart.read()                            # Read data
     if buf:
      dat = buf.decode('UTF-8')                           # Decode
      if dat[0] == 'R' and dat[1] == '1':         # R1?
         R.value(1)                                # LED ON
         uart.write("R LED is ON\n")               # Send confirmation
      elif dat[0] == 'R' and dat[1] == '0':       # R0?
         R.value(0)                               # LED OFF
         uart.write("R LED is OFF\n")              # Send confirmation

      elif dat[0] == 'G' and dat[1] == '1':       # G1?
         G.value(1)                               # LED OFF
         uart.write("G LED is ON\n")               # Send confirmation
      elif dat[0] == 'G' and dat[1] == '0':       # G0?
         G.value(0)                               # LED OFF
         uart.write("G LED is OFF\n")              # Send confirmation

      elif dat[0] == 'B' and dat[1] == '1':       # B1?
         B.value(1)                               # LED OFF
         uart.write("B LED is ON\n")               # Send confirmation
      elif dat[0] == 'B' and dat[1] == '0':       # B0?
         B.value(0)                               # LED OFF
         uart.write("B LED is OFF\n")              # Send confirmation
```

Figure 10.4: Program: **BlueLED***.*

Testing the Program

The program can be tested by using an Android mobile phone to send commands through a Bluetooth communication interface. There are many freely available Bluetooth communication programs in the Play Store. The one chosen by the author was called the **Bluetooth Controller** by *mightyIT* (it@memighty.com) as shown in Figure 10.5. You should install this app on your Android mobile phone so that you can send commands to the development board.

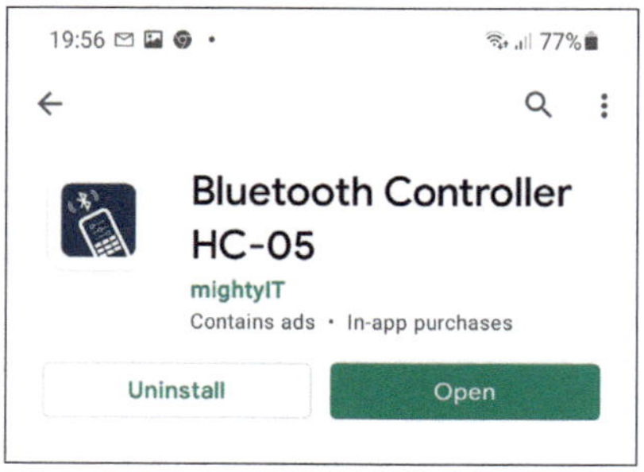

Figure 10.5: Bluetooth Controller apps.

The steps to test the application are as follows:

- Construct the project.
- Download the program to your Raspberry Pi Pico W and start it.
- Activate the **Bluetooth Controller** app on your mobile phone.
- The apps will look for nearby Bluetooth devices. Click on **HC-06** when displayed on the mobile phone screen (you may have to scan for devices).
- You may now be asked to enter the password to pair the mobile phone with the Pico. Enter the default password of **1234**.
- Click the semicircle with an arrow located at the top right side of the screen to connect to HC-06.
- You should see a green colour dot at the top right-hand side of the screen when a connection is made to HC-06. Also, HC-06 with its address (e.g., HC-06 [98:D3:91:F9:6C:19] should be displayed at the top left side of the screen.
- To turn the R LED ON, enter command **R1** and click **Send ASCII**. You should see the R LED turning ON. Enter command **R0** to turn OFF the LED. Figure 10.6 shows an example screen.

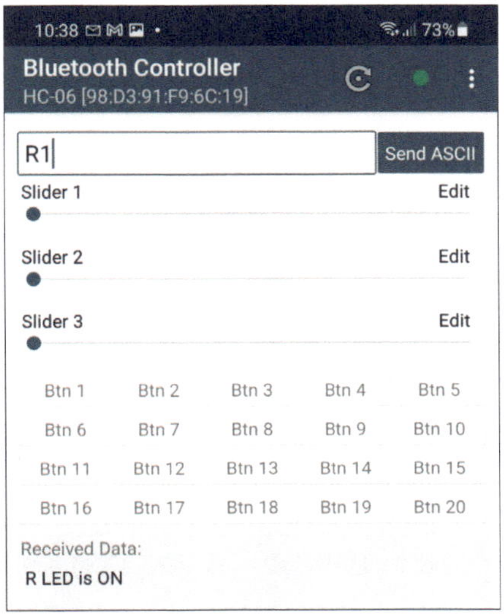

Figure 10.6: Example command to turn the R LED ON.

Figure 10.7 shows the project built on a breadboard.

Figure 10.7: Project built on a breadboard.

10.4 Project 2: Sending the Raspberry Pi Pico W internal temperature to a smartphone

Description: In this project, the internal temperature of the Raspberry Pi Pico is read every 10 seconds and is sent to a smartphone over a Bluetooth link.

Aim: The aim of this project is to show how data can be sent from the Raspberry Pi Pico to a smartphone at regular intervals

Block Diagram: Figure 10.8 shows the block diagram of the project.

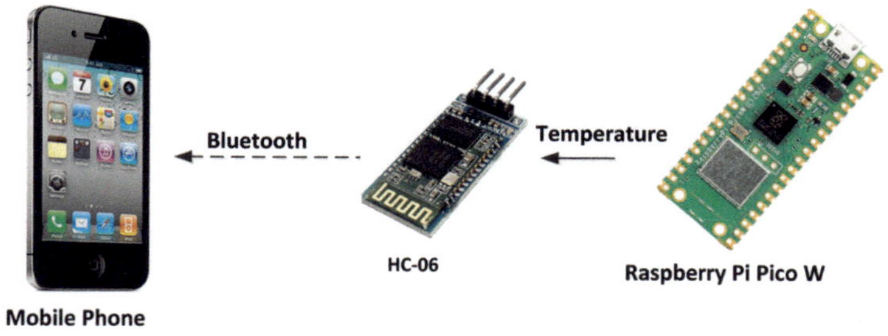

Figure 10.8: Block diagram of the project.

Circuit Diagram: Figure 10.9 shows the project circuit diagram. RXD and TXD pins of the Bluetooth module are connected to UART 0 pins GP0 (TX) and GP1 (RX) of Raspberry Pi Pico W, respectively.

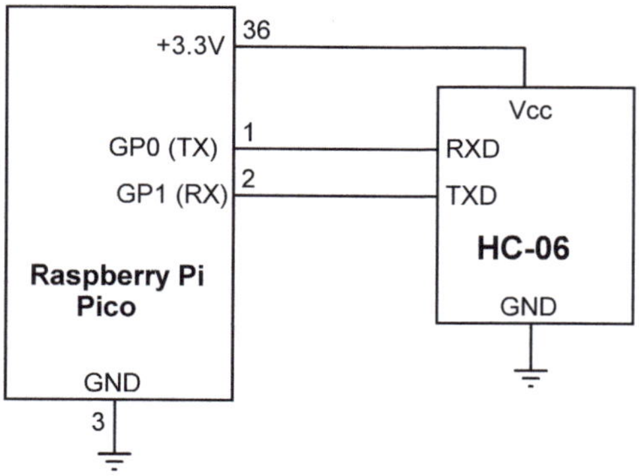

Figure 10.9: Circuit diagram of the project.

Program listing: The program listing of the project is shown in Figure 10.10 (program: **BlueTemp**). At the beginning of the program, the hardware UART interface is set to a baud rate of 9600 which is the default speed of HC-06. Function **GetTemperature** returns the internal temperature of the Raspberry Pi Pico W. Inside the main program loop, the temperature is read and sent to the smartphone every 10 seconds. Figure 10.11 shows the temperature displayed on a smartphone using the **Bluetooth Controller** apps described in the previous section.

```
#------------------------------------------------------------
#               SEND TEMPERATURE TO SMART PHONE
#               ===============================
#
# In this project a HC-06 type serial Bluetooth module is
# connected to the Raspberry Pi Pico. Internal temperature
# readings are sent to a smart phone every 10 seconds
#
```

```
# Author: Dogan Ibrahim
# File  : BlueTemp.py
# Date  : October 2022
#--------------------------------------------------------------
from machine import Pin, UART, ADC
import utime

uart = UART(0, baudrate=9600,rx=Pin(1),tx=Pin(0))

Conv = 3.3 / 65535
AnalogIn = ADC(4)

def GetTemperature():
    V = AnalogIn.read_u16()
    V = V * Conv
    Temp = 27 - (V - 0.706) / 0.001721
    return Temp

#
# Send the temperature to smart phone
#
while True:
     T = GetTemperature()
     Temp = "T=" + str(T)[:5] + "\r\n"
     uart.write(Temp)
     utime.sleep(10)
```

Figure 10.10: Program: **BlueTemp***.*

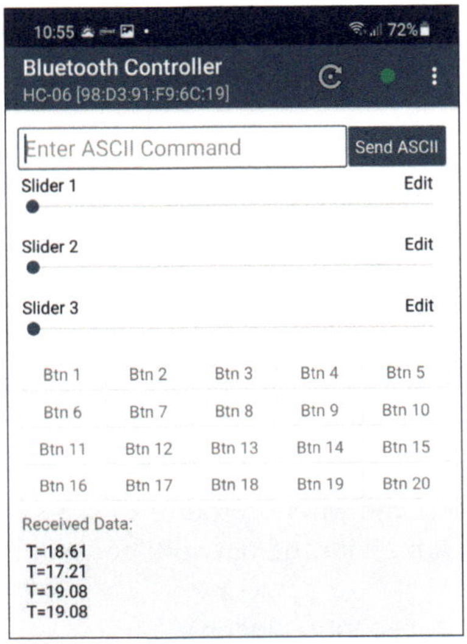

Figure 10.11: Temperature displayed on the smartphone.

Chapter 11 • Using Wi-Fi with the Raspberry Pi Pico W

11.1 Overview

The Raspberry Pi Pico W adds Wi-Fi capability to the Raspberry Pi Pico, allowing you to connect the device to a Wi-Fi network. The new Pico W adds a 2.4 GHz 802,11b/g/n Infineon CYW43439 Wi-Fi chipset with on-board antenna. This chipset supports both Wi-Fi, Bluetooth, and Bluetooth Low Energy (BLE). At the time of writing this book, Bluetooth firmware was not available.

Perhaps the first thing you may want is to try to connect to a wireless network. As shown in this chapter, you will need to create a WLAN object, use its connect method to connect an SSID and password. You should be given an IP address. You can then develop Wi-Fi-based programs using the Micropython programming language.

11.2 Connecting to a wireless network

The steps to connect to a wireless network are described below with function connect():

```
import machine
import network
import utime

def connect():
wlan = network.WLAN(network.STA_IF)
wlan.active(True)
wlan.connect("MySSID", "Mypassword")
while wlan.isconnected() == False:
            print("Waiting to be connected")
            utime.sleep(1)
```

After connecting, you can display the IP of your connection with the following statement:

```
print(wlan.ifconfig())
```

11.3 Project 1: Scan the local network

Description: This is a simple Wi-Fi example that scans the local Wi-Fi network router and displays the users of this router on the Thonny screen.

Aim: The aim of this project is to show how to scan the local network.

Program listing: Figure 11.1 shows the program listing (Program: **scanwifi.py**). Notice that there is no need to connect the wireless router to scan the network. Function **scan()** returns the following information about the network nodes it finds:

(ssid, bssid, channel, RSSI, security, hidden)

In this program, only the SSID is displayed.

bssid is the hardware address of an access point, in binary form, returned as bytes object. You can use **binascii.hexlify()** to convert it to ASCII form.

There are five values for security:

- 0 – open
- 1 – WEP
- 2 – WPA-PSK
- 3 – WPA2-PSK
- 4 – WPA/WPA2-PSK

and two for hidden:

- 0 – visible
- 1 – hidden

Some other available network functions are:

wlan.disconnect(): Disconnect from network
wlan.status() Return current status of the wireless connection
wlan,config(): Get/set general network parameters

Further information about the network functions can be found at the following link:

https://docs.micropython.org/en/latest/library/network.WLAN.html

```
#-----------------------------------------------------------
#              SCAN LOCAL WIRELESS NETWORK
#              ===========================
#
# This program scans the local wireless network and displays
# the users on this network
#
# Author: Dogan Ibrahim
# File  : scanwifi.py
# Date  : October 2022
#-------------------------------------------------------------
import machine
import network
import utime

def connect():
    global wlan
    wlan = network.WLAN(network.STA_IF)
```

```
    wlan.active(True)
#    wlan.connect("BTHomeSpot-XNH", "49345axwub")
#    while wlan.isconnected() == False:
#                        print("Waiting to be connected")
#                        utime.sleep(1)
#    print("Connected")

connect()
wifis = wlan.scan()

i=0
for w in wifis:
      i+=1
      print(i,w[0].decode())
```

Figure 11.1: Program: **scanwifi.py**.

Figure 11.2 shows the data displayed on the Thonny screen.

```
1 The Hotal
2 BTHub5-6SPN
3 DIRECT-1F-HP Laser 150nw
4 BTHomeSpot-XNH
5
6 BTWi-fi
7 BTB-KNCGPZ
8 Family room TV.b,
9 BTHomeSpot-XNH
10 BTWi-fi
11
12 World Machine
```

Figure 11.2: Data displayed.

You may want to close the connection using the **wlan.disconnect()** function after finishing your work.

11.4 Using the Socket library

The Socket library is mainly used in network-based communications. UDP and TCP/IP are the two most commonly used protocols to send and receive data over a network. TCP/IP is a reliable protocol that includes handshaking and thus guarantees the delivery of packets to the required destination. UDP, on the other hand, is not so reliable but is a fast protocol.

Table 11.1 shows a comparison of the UDP and TCP/IP type communications.

TCP/IP	UDP
Connection-oriented protocol	Connectionless protocol
Slow	Fast
Highly reliable data transmission	Not so reliable
Packets arranged in order	No ordering of packets
20-byte header size	8-byte header size
Error checking and re-transmission	No error checking
Data received acknowledgement	No acknowledgement
Used in HTTP, HTTPS, FTP etc.	Used in DNS, DHCP, TFTP etc.

Table 11.1: Comparison of UDP and TCP/IP.

UDP and TCP/IP protocol-based programs are server-client based where one node sends data, and the other node receives it and vice-versa. Data is transferred through ports where the server and the clients must use the same port numbers.

In the next sections, you shall be looking at how UDP and TCP/IP protocol-based programs can be written using MicroPython for the ESP32 processor.

11.4.1 UDP programs

UDP is a connectionless protocol and thus there is no need to make a connection to the destination node before a packet can be sent to it. The communication between a server and a client is basically as follows:

Server
Define the destination node IP address and port number.
Construct the data packet.
Create a socket.
Bind the socket to the local port.
Receive data from the client.
Send data to the client.
Close the connection.

Client
Define the destination node IP address and port number.
Create a socket.
Send data to the server.
Receive data from the server.
Close the connection.

Notice that both the server and the client can send and receive data packets from each other.

11.5 Project 2: Controlling an LED from a smartphone using Wi-Fi – UDP communication

Description: In this project, you will be sending commands over the Wi-Fi link from a smartphone to control an LED (the LED can be replaced with a relay for example to control an equipment) connected to the Raspberry Pi Pico W. Valid commands are (the command must be terminated with a newline):

LON	Turn LED ON
LOFF	Turn LED OFF

Aim: The aim of this project is to show how to use the Wi-Fi connectivity on the Raspberry Pi Pico W.

Block Diagram: Figure 11.3 shows the project block diagram.

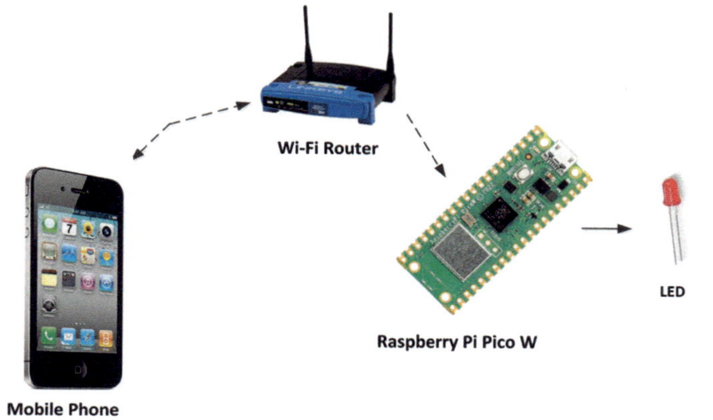

Figure 11.3: Block diagram of the project.

The LED is connected to GP15 (pin 20) of the Raspberry Pi Pico W through a 470 Ohm current limiting resistor.

Program Listing: Before writing the program, you should find the IP address assigned to your Pico. This can be displayed on the Thonny screen by including the following statement: **print(wlan.ifconfig())** after connecting to the Wi-Fi. In this project, the IP address of the Pico was: 192.168.1.154.

Figure 11.4 shows the program listing (program: **Picowifi**). Function **Connect()** is called to connect to the local Wi-Fi router. The remaining of the program runs in an endless loop formed using a **while** statement. Inside this loop, data is received from the smartphone over a UDP link with port 5000, and the LED is controlled accordingly. Commands **LON** and **LOFF** turn the LED ON and OFF respectively.

```
#-----------------------------------------------------------
#              REMOTE WI-FI BASED LED CONTROL
#              ==============================
#
# In this program an LED (or a buzzer,relay etc) connected
# to Pico is controlled from a smartphone over a wi-fi link
#
# Author: Dogan Ibrahim
# File  : picowifi.py
# Date  : October 2022
#-------------------------------------------------------------
from machine import Pin
import network
import socket
import utime

LED = Pin(15, Pin.OUT)                  # LED at GP15
LED.value(0)                            # LED OFF

#
# This function attempts to connect to Wi-Fi
#
def connect():
    global wlan
    wlan = network.WLAN(network.STA_IF)
    wlan.active(True)
    wlan.connect("BTHomeSpot-XNH", "4933axb")
    while wlan.isconnected() == False:
               print("Waiting to be connected")
               utime.sleep(1)
    print("Connected")

connect()
port = 5000
UDP = ("", port)
sock = socket.socket(socket.AF_INET, socket.SOCK_DGRAM)
sock.bind(UDP)

#
# Main program loop
#
while True:
     buf, addr = sock.recvfrom(1024)
     if buf:
        buf = buf.decode('utf-8')
        if buf[0]=="L" and buf[1]=="O" and buf[2]=='N':
```

```
    LED.value(1)
  if buf[0]=="L" and buf[1]=='O' and buf[2]=='F' and buf[3]=='F':
    LED.value(0)
```

Figure 11.4: Program: **Picowifi***.*

Testing the program

The program can easily be tested using the **PacketSender** program (see Figure 11.5) on the PC or using a smartphone after installing a UDP app.

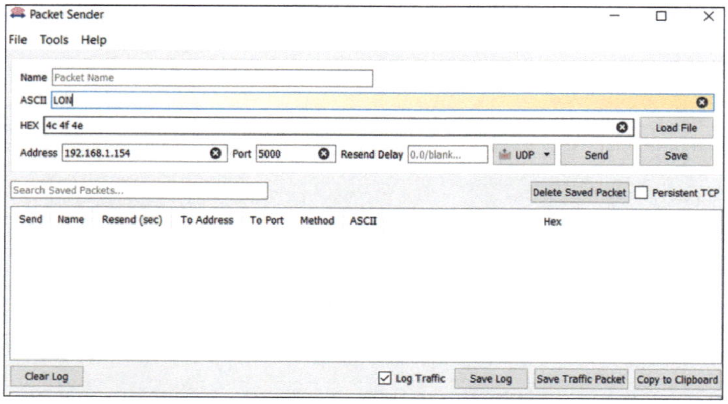

Figure 11.5: Using the PacketSender to test the program.

You should install a **UDP Server** app on your Android mobile phone before starting the test with the smartphone. There are many freely available UDP apps in the **Play Store**. The one installed and used in this project is called the **UDP Terminal** by *mightyIT* as shown in Figure 11.6.

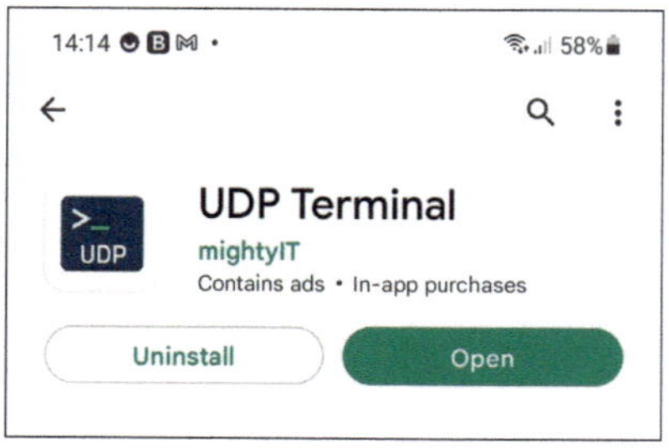

Figure 11.6: UDP Terminal.

The steps to test the program are as follows:

- Construct the circuit.
- Download the program to your Raspberry Pi Pico W and start the program.
- Start the **UDP Terminal** apps on your smartphone.
- Set the Pico IP address (192.168.1.154 in this example) as shown in Figure 11.7.

- Click **Start Terminal.**
- Enter command **LON** at the bottom of the screen to turn ON the LED and click **Send ASCII.** The LED should turn ON (Figure 11.8).

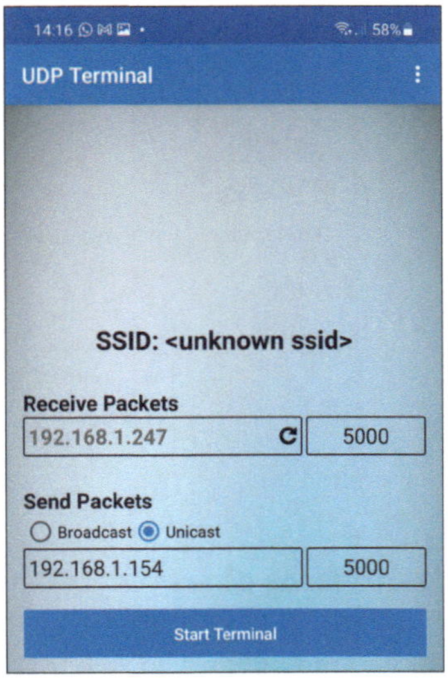

Figure 11.7: Start the UDP terminal.

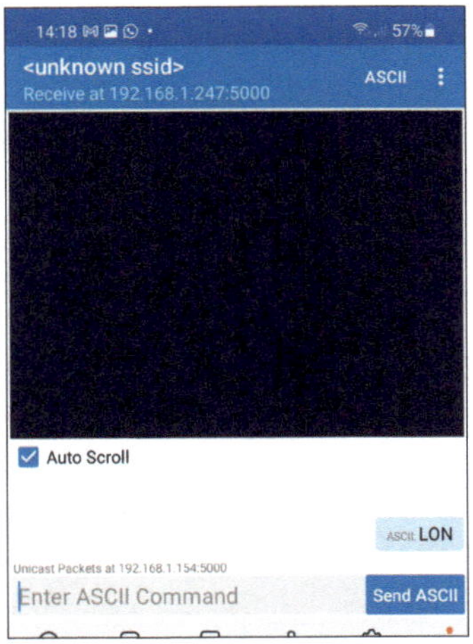

Figure 11.8: Click Send Ascii.

11.6 Project 3: Displaying the internal temperature on a smartphone using Wi-Fi

Description: In this project, you will be reading the internal temperature of the Raspberry Pi Pico W and then send this data to a smartphone over a Wi-Fi link.

A request for data is made by the smartphone (e.g., **T?**) when it requires data from the Pico. This project uses two-way UDP communications to receive and send data.

Aim: The aim of this project is to show how two-way communication can be established with a smartphone over the Wi-Fi link.

Block Diagram: Figure 11.9 shows the block diagram of the project.

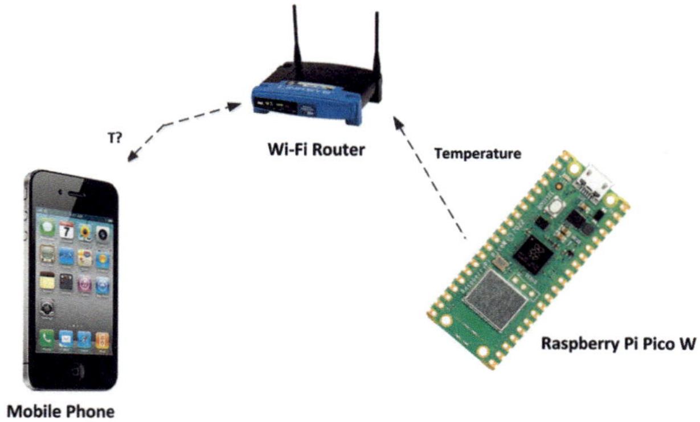

Figure 11.9: Block diagram of the project.

Program Listing: The program listing is shown in Figure 11.10 (program: **temptowifi**). Function **Connect()** is called to connect to the local Wi-Fi router as in the previous program. Inside the main program loop, the program waits to receive command **T?** and when this command is received, function **GetTemperature** is called and the Raspberry Pi Pico W internal temperature is read and stored in variable **T** in the main program. String variable **Tstr** stores string **T=** followed by the value of the temperature as a string. The length of this string is stored in variable **Tlen** and is then sent to the smartphone using the network **sendto()** function.

```
#------------------------------------------------------------
#               SEND TEMPERATURE TO SMART PHONE
#               ===============================
#
# In this project internal temperature of the Raspberry Pi
# Pico W is sent to the smartphone
#
# Author: Dogan Ibrahim
# File  : temptowifi.py
# Date  : October 2022
#--------------------------------------------------------------
from machine import ADC
import network
import socket
import utime

AnalogIn = ADC(4)                      # ADC channel 4
Conv = 3300 / 65535                    # Conversion factor
```

```
Conv = 3.3 / 65535                    # Conversion factor

#
# This function returns the internal temperature
#
def GetTemperature():
    V = AnalogIn.read_u16()
    V = V * Conv
    Temp = 27 - (V - 0.706) / 0.001721
    return Temp

#
# This function attempts to connect to Wi-Fi
#
def connect():
    global wlan
    wlan = network.WLAN(network.STA_IF)
    wlan.active(True)
    wlan.connect("BTHomeSpot-XNH", "49345xsdaez")
    while wlan.isconnected() == False:
            print("Waiting to be connected")
            utime.sleep(1)
    print("Connected")

connect()
port = 5000
UDP = ("", port)
sock = socket.socket(socket.AF_INET, socket.SOCK_DGRAM)
sock.bind(UDP)

#
# Main program loop
#
while True:
     buf, addr = sock.recvfrom(1024)
     if buf:
        buf = buf.decode('utf-8')
        if buf[0]=="T" and buf[1]== '?':
           T = GetTemperature()
           Tstr = "T=" + str(T)[:5]
           Tlen = len(Tstr)
           sock.sendto(Tstr, addr)
```

Figure 11.10: Program: **temptowifi**.

Testing the program

The program can either be tested on a PC using the freely available **PacketSender** program or a smartphone with a UDP app installed. **PacketSender** enables the user to send and receive UDP as well as TCP packets from a PC. This program could be very useful during the testing of UDP- and TCP-based programs.

Alternatively, you can use the **UDP Terminal** app as in the previous project to test the program. An example run of the program on the apps is shown in Figure 11.11, where command **T?** is sent to the Raspberry Pi Pico W and the internal temperature is extracted and displayed on the Android screen. Note here that the IP address of the Pico was: 192.168.1.154.

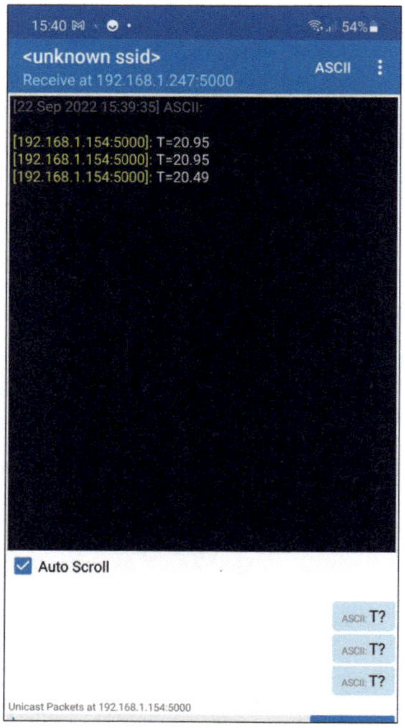

Figure 11.11: Example run of the program.

11.7 Project 4: Remote control from an Internet browser (using a smartphone or PC) - Web Server

Description: In this project, two LEDs are connected to Raspberry Pi Pico W and are controlled remotely from a device that can run an Internet browser (e.g., Firefox, etc.). The idea here is to control the LEDs remotely from a web link. In practical applications, these LEDs can be replaced by relays and the project can be used to control any kind of electrical equipment connected to the development board.

Valid commands are as follows (any other commands are ignored by the program):

0=ON	Turn ON LED0
0=OFF	Turn OFF LED0
1=ON	Turn ON LED1
1=OFF	Turn OFF LED1

Block diagram: Figure 11.12 shows the block diagram of the project.

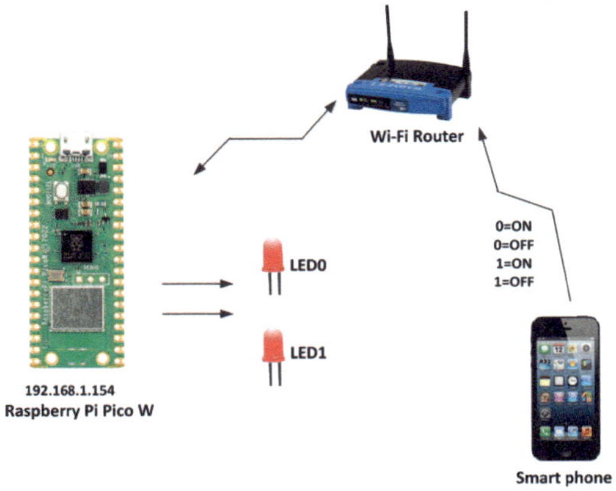

Figure 11.12: Block diagram of the project.

Circuit diagram: The circuit diagram of the project is shown in Figure 11.13. LED0 and LED1 are connected to port pins GP14 (pin 19) and GP15 (pin 20) through 470-Ohm current-limiting resistors, respectively.

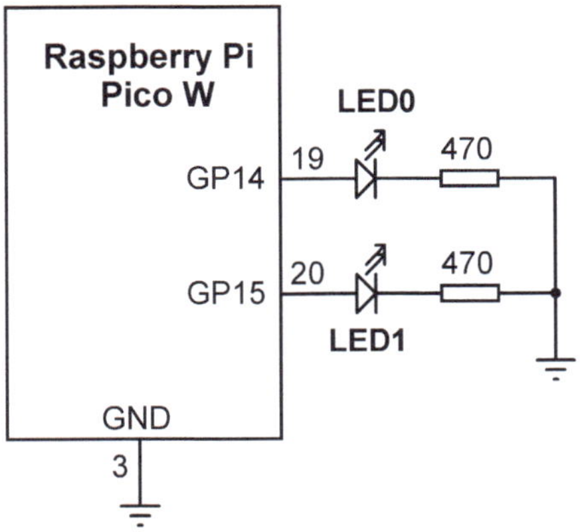

Figure 11.13: Circuit diagram of the project.

Program listing: Figure 11.14 shows the program listing (Program: **WebServer.py**). At the beginning of the program, the network and socket libraries are imported into the program. Then, GP14 and GP15 are configured as outputs. The program then defines the HTML code that is to be sent to the web client when a request comes from it. This code is stored in variable **html**. The web server program sends the html file (statements **resp=html** and **conn.send(resp))**) to the client and waits for a command from the client. When a button is clicked on the client, a command is sent to the web server in addition to some other data. Here, you are only interested in the actual command sent. The following commands are sent to the web server for each button pressed:

Button Pressed	**Command sent to web server**
LED0 ON	/?LED0=ON
LED0 OFF	/?LED0=OFF
LED2 ON	/?LED2=ON
LED2 OFF	/?LED2=OFF

The web server program on Pico makes a TCP connection and reads the command from the web client (**request = conn.recv(1024)**). The program then searches the received command to see if any of the above commands are present in the received data. Remember that **find** searches a string for a sub-string and returns the starting position of the sub-string in the string. If the sub-string is not found, then a "-1" is returned. For example, consider the following statement:

LEDON0 = request.find("/?LED0=ON")

Here, sub-string **/?LED0=ON** is searched in the string **request** and if found, the starting position of the sub-string in the string is given. A "-1" is returned if the sub-string **/?LED0=ON** does not exist in string **request**. The program looks for a match for all four commands. If a command is found, then the corresponding LED is turned ON or OFF accordingly:

```
LEDON0 = request.find("/?LED0=ON")
       LEDOFF0 = request.find("/?LED0=OFF")
       LEDON2 = request.find("/?LED2=ON")
       LEDOFF2 = request.find("/?LED2=OFF")
       if LEDON0 != -1:                                  # If ON0 command found
                       LED0.value(1)                     # Turn ON LED0
       if LEDOFF0 != -1:                                 # If OFF0 command found
                       LED0.value(0)                     # Turn OFF LED0
        if LEDON2 != -1:                                 # If ON2 command found
                       LED2.value(1)                     # Turn ON LED2
       if LEDOFF2 != -1:                                 # If OFF2 command found
                       LED2.value(0)                     # Turn OFF LED2
```

In this project, the IP address of the web server was 192.168.1.154. The procedure to test the system is as follows:

- Run the MicroPython program in Figure 11.14 on the Pico.
- Open your web browser on your PC (or any other device with an Internet browser) and enter the web site address 192.168.1.154.

- The control HTML form will be displayed on your PC screen as shown in Figure 11.15. Click the buttons to turn ON/OFF the required LED.

```
#------------------------------------------------------------------------------
#                      WEB SERVER TO CONTROL 2 LEDs
#                      ============================
#
# This is a simple web server program. 2 LEDs are conenected to GP14
# and GP15 of Raspberry Pi Pico W. The project controls these LEDs
# (turns ON or OFF) from a web server application. For example, the
# LEDs can be controlled from any device that is on the web, for example
# a PC, tablet, mobile phone etc. When activated, a form will appear on
# the device with buttons and clicking these button will control the LEDs.
# The LEDs are named as LED0 and LED1
#
# Author: Dogan Ibrahim
# Date  : October, 2022
# File  : WebServer.py
#-------------------------------------------------------------------------------
from machine import Pin
import network
import socket
import utime

#
# This function attempts to connect to Wi-Fi
#
def connect():
    global wlan
    wlan = network.WLAN(network.STA_IF)
    wlan.active(True)
    wlan.connect("BTHomeSpot-XNH", "49345azse5")
    while wlan.isconnected() == False:
               print("Waiting to be connected")
               utime.sleep(1)
    print("Connected")

#
# Configure the pins as outputs
#
LED0 = Pin(14, Pin.OUT)
LED1 = Pin(15, Pin.OUT)
#
# Turn OFF the LEDs to start with
#
LED0.value(0)
```

```
LED1.value(0)

#
# The HTML code
#
html = """<!DOCTYPE html>
<html>
<body>
<h1>Raspberry Pi Pico W LED ON/OFF</h1>
<h2>MicroPython Web Server Example with 2 LEDs</h2>
<form>

<button name="LED0" button style="color:green" value="ON" type="submit">LED0 ON</
button>
<button name="LED0" button style="color=red" value="OFF" type="submit">LED0 OFF</
button><br><br>

<button name="LED1" button style="color:green" value="ON" type="submit">LED1 ON</
button>
<button name="LED1" button style="color:red" value="OFF" type="submit">LED1 OFF</
button>
</form>
</body>
</html>
"""

#
# Connect to local Wi-Fi
#
connect()
addr=socket.getaddrinfo("0.0.0.0",80)[0][-1]
s = socket.socket()
s.bind(addr)
s.listen(1)

while True:
    conn, address = s.accept()
    request = conn.recv(1024)
    request = str(request)
    LEDON0 = request.find("/?LED0=ON")
    LEDOFF0 = request.find("/?LED0=OFF")
    LEDON1 = request.find("/?LED1=ON")
    LEDOFF1 = request.find("/?LED1=OFF")
    if LEDON0 != -1:
        LED0.value(1)
    if LEDOFF0 != -1:
```

```
    LED0.value(0)
if LEDON1 != -1:
    LED1.value(1)
if LEDOFF1 != -1:
    LED1.value(0)
resp = html
conn.send(resp)
conn.close()
```

Figure 11.14: Program: **WebServer.py**.

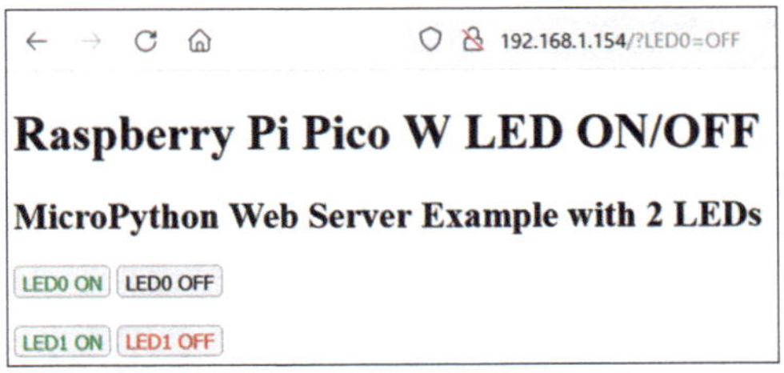

Figure 11.15: Form displayed on the PC.

11.8 Project 5: Storing ambient temperature and atmospheric pressure data on the Cloud

Description: In this project, the ambient temperature and atmospheric pressure are read and stored on the Cloud. A BMP280-type sensor module is used in this project.

BMP280

The BMP280 is an absolute barometric pressure and temperature sensor chip, which has been developed for mobile applications. Its small dimensions and its low power consumption allow for the implementation in battery-powered devices such as mobile phones, GPS modules or watches. The BMP280 is based on Bosch's proven piezo-resistive pressure sensor technology featuring high accuracy and linearity as well as long-term stability and high EMC robustness. The device is optimized in terms of power consumption, resolution, and filter performance. The basic specifications of the BMP280 are:

- Pressure range: 300 to 1100 hPa
- Relative accuracy: ±0.12 hPa
- Absolute accuracy: ±1 hPa
- Digital interface: I2C, SPI
- Temperature range: -40 ^{o}C to +85 ^{o}C
- Current consumption: 2.7 µA

Block diagram: The block diagram of the project is shown in Figure 11.16.

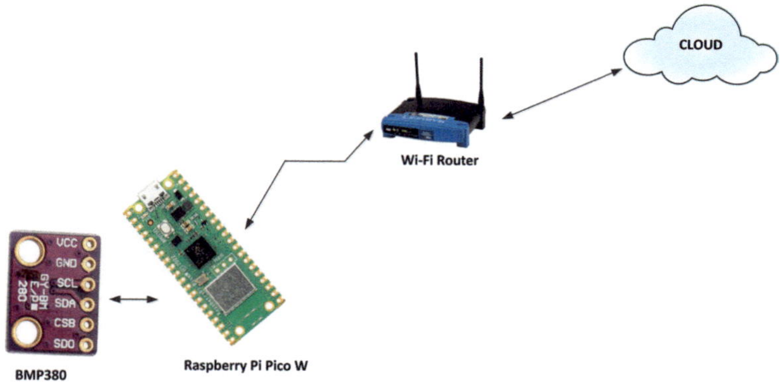

Figure 11.16: Block diagram of the project.

Circuit diagram: Figure 11.17 shows the circuit diagram. SCL and SDA pins of BMP280 are connected to SCL (pin 2) and SDA (pin 1) of Pico. The sensor is powered from +3.3 V.

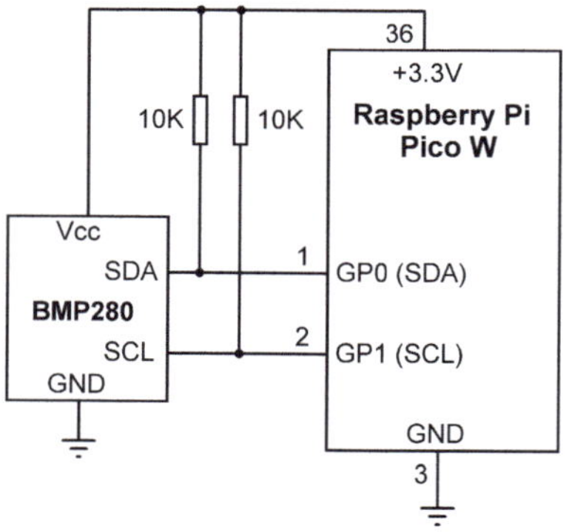

Figure 11.17: Circuit diagram of the project.

The Cloud

There are several cloud services that can be used to store data (for example **SparkFun, ThingSpeak, Cloudino, Bluemix**, etc.,). In this project, the **Thingspeak** is used. This is a free cloud service where sensor data can be stored and retrieved using simple HTTP requests. Before using the Thingspeak, you have to create an account from their web site and then log in to this account.

Go to the ThingSpeak web site:

https://thingspeak.com/

*Click **Get Started For Free** and create an account if you don't already have one*. Then, you should create a New Channel by clicking on **New Channel**. Fill in the form as shown in Figure 11.18. Give the name **Raspberry Pi Pico W** to the application, give a description, and create two **fields** called **Atmospheric Pressure** and **Temperature**. You can optionally fill in the other items as well if you wish.

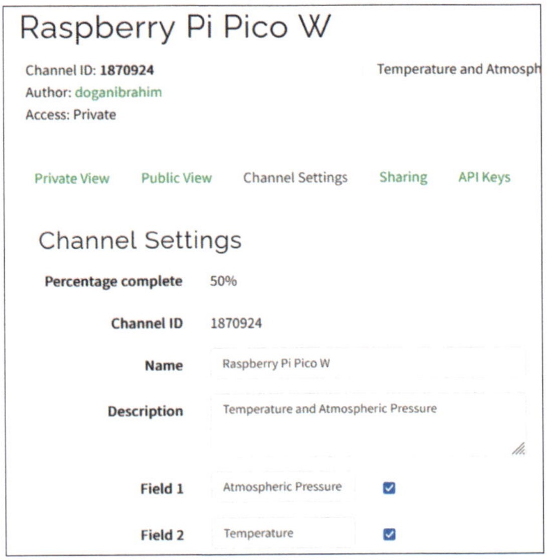

Figure 11.18: Create a New Channel (only part of the form shown).

Click **Save Channel** at the bottom of the form. Your channel is now ready to be used with your data. You will now see tabs with the following names. You can click on these tabs and see the contents to make corrections if necessary:

- **Private View**: This tab displays private information about your channel where only you can see.
- **Public View**: If your channel is public, use this tab to display selected fields and channel visualizations.
- **Channel Settings**: This tab shows all the channel options you set at creation. You can edit, clear, or delete the channel from this tab.
- **API Keys**: This tab displays your channel API keys. Use the keys to read from and write to your channel.
- **Data Import/Export**: This tab enables you to import and export channel data.

You should click the **API Keys** tab and save your **Write API** and **Read API** keys and the **Channel ID** in a safe place as you will need to use them in your program. The API Keys and the Channel ID in this project were:

Also, note your **Channel Number**: 1870924

Also, select the **Public View** and navigate to **Sharing**. You may select option **Share channel view with everyone** so that everyone can access your data remotely.

Program listing: Figure 11.19 shows the program listing (Program: **Cloud.py**). Before using the software, you have to upload the BMP280 library to your Pico. This library is included at the web site of the book under the name **bmp280.py**. You should copy this file to your Raspberry Pi Pico W with the exact name **bmp280.py**.

At the beginning of the program, the network libraries, time library, bmp280 library and the I²C library are imported to the program. Thingspeak **Write Key** and **Host Address** are defined. Function **connect()** connects to the local Wi-Fi. The main program loop starts with the **while** statement. Inside this loop, the IP address of the **Thingspeak** web site is extracted and a connection is made to this site at port 80. Then the atmospheric pressure and temperature readings are obtained from the BMP280 module and included in the **path** statement. The **sock.send** statement sends an HTTP GET request to the **ThingSpeak** site and uploads the pressure and temperature values. This process is repeated every 30 seconds.

Figure 11.20 shows the pressure and temperature data plotted by **ThingSpeak**. The Chart Options can be clicked to change various parameters of the charts. For example, Figure 11.21 shows the temperature as a column display. In Figure 11.22 the pressure is shown as a step graph. A title and X-axis label is added in Figure 11.23 to the pressure graph. Figure 11.24 shows the atmospheric pressure displayed in a clock format (click **Add Widgets** for this type of display).

Because the Channel was saved as Public, you can view the graph from a web browser by entering the Channel ID. In this project, the link to view the data graphs from a web browser is:

https://api.thingspeak.com/channels/1870924

Figure 11.25 shows the graphs accessed from a web browser. You can also export some or all the fields in CSV format by clicking **Export recent data**, so that it can be analyzed by external statistical packages such as Excel (Figure 11.26).

```
#----------------------------------------------------------------------------
#        ATMOSPHERIC PRESSURE AND TEMPERATURE ON THE CLOUD
#       ================================================
#
# The ambient temperature and pressure sensor BMP280 is connected to Pico.
# The project reads the temperature and atmospheric pressure and sends
# to the cloud where it can be accessed from anywhere. In addition, change
# of the temperature and the pressure can be plotted in the cloud.
#
#
# The program uses the Thingspeak cloud service
#
# Author: Dogan Ibrahim
# File  : Cloud.py
# Date  : January, 2021
#-----------------------------------------------------------------------------
import network
import socket
import utime
from machine import Pin,I2C
from bmp280 import *

bus = I2C(0,scl=Pin(1),sda=Pin(0),freq=200000)
bmp = BMP280(bus)

bmp.use_case(BMP280_CASE_INDOOR)
APIKEY = "BUHXZZC718WT0X0P"                    # Thingspeak API key
host = "api.thingspeak.com"                    # Thigspeak host

#
# This function attempts to connect to Wi-Fi
#
def connect():
    global wlan
    wlan = network.WLAN(network.STA_IF)
    wlan.active(True)
    wlan.connect("BTHomeSpot-XNH", "49345we34g")
    while wlan.isconnected() == False:
            print("Waiting to be connected")
            utime.sleep(1)


#
# Send data to Thingspeak. This function sends the temperature and
# humidity data to the cloud every 30 seconds
#
connect()
```

```
while True:
    connect()
    sock = socket.socket()
    addr = socket.getaddrinfo("api.thingspeak.com",80)[0][-1]
    sock.connect(addr)
    pressure=bmp.pressure
    p=pressure/133.3224                    # Pressure in mmH
    t=bmp.temperature                      # Temperature in C
    path = "api_key="+APIKEY+"&field1="+str(p)+"&field2="+str(t)
    sock.send(bytes("GET /update?%s HTTP/1.0\r\nHost: %s\r\n\r\n"
%(path,host),"utf8"))
    utime.sleep(5)
    sock.close()
    wlan.disconnect()
    utime.sleep(25)
```

Figure 11.19: Program: **Cloud.py**.

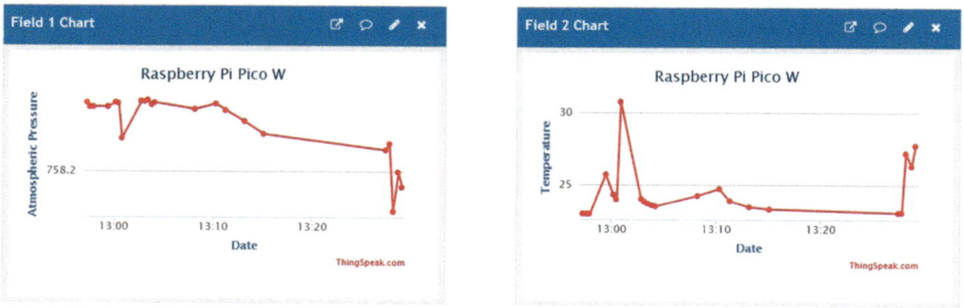

Figure 11.20: Plotting the pressure and temperature.

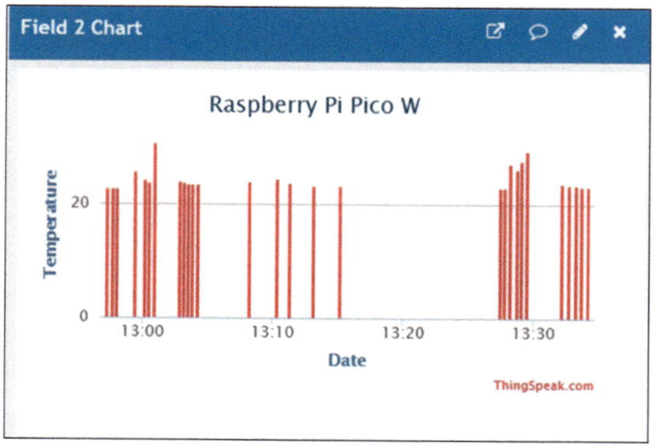

Figure 11.21: Displaying temperature as columns.

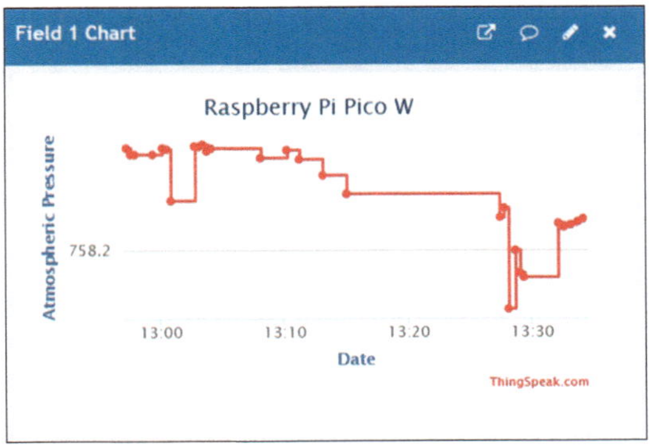

Figure 11.22: Displaying pressure as steps.

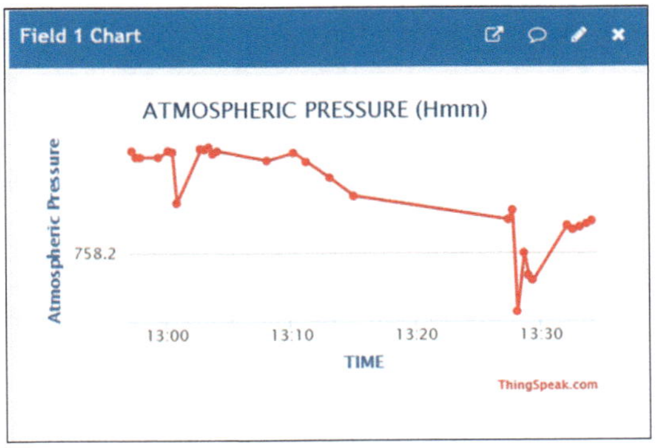

Figure 11.23: Adding title and x-axis label.

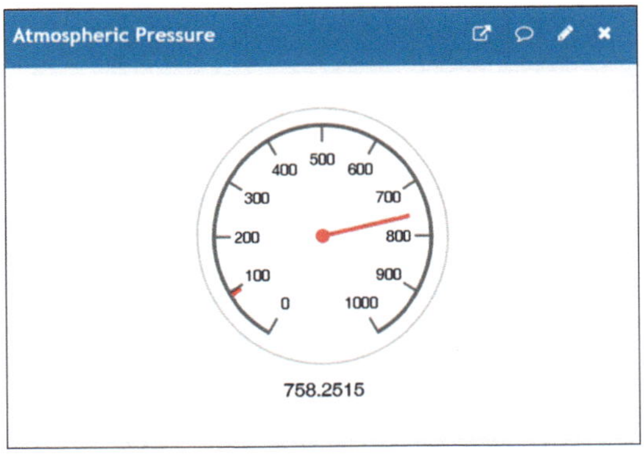

Figure 11.24: Displaying pressure in a clock format.

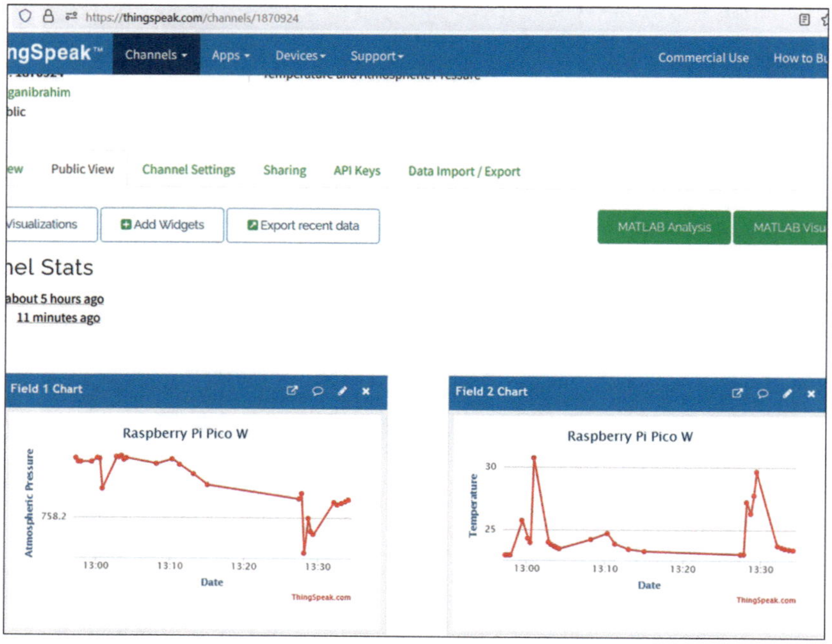

Figure 11.25: Graph accessed from a web server.

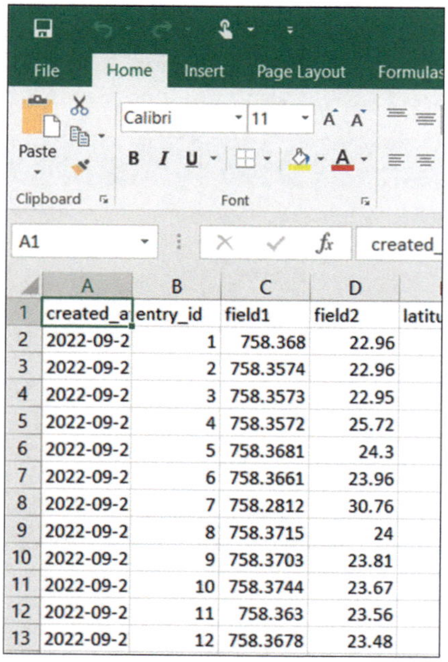

	A	B	C	D	
1	created_a	entry_id	field1	field2	latitu
2	2022-09-2	1	758.368	22.96	
3	2022-09-2	2	758.3574	22.96	
4	2022-09-2	3	758.3573	22.95	
5	2022-09-2	4	758.3572	25.72	
6	2022-09-2	5	758.3681	24.3	
7	2022-09-2	6	758.3661	23.96	
8	2022-09-2	7	758.2812	30.76	
9	2022-09-2	8	758.3715	24	
10	2022-09-2	9	758.3703	23.81	
11	2022-09-2	10	758.3744	23.67	
12	2022-09-2	11	758.363	23.56	
13	2022-09-2	12	758.3678	23.48	

Figure 11.26: Exporting data in Excel format.

Chapter 12 • RFID Projects

12.1 Overview

RFID is a Radio Frequency Identification device used for security and tracking purposes. An RFID system includes a reader card and a tag, and both of them are included with the kit. RFID uses electromagnetic fields to transfer data over short distances. RFID systems are mainly used in security applications, for example, they can be used to open a door where the person having the right tag is allowed to open the door.

One of the popular RFID readers is the RC522 module (Figure 12.1) with the following basic specifications:

- Operating frequency: 13.56 MHz
- Operating voltage: +2.5 V to +3.3 V
- Operating current: 13 to 26 mA
- Read range: 5 cm
- Operation with both SPI bus and $I^{2C\ bus}$

Figure 12.1: RFID reader and tag.

The RFID reader is usually supplied with header pins which must be soldered to the sockets at the edge of the reader before it can be used. The RC522 RFID module is supplied with RFID card tag and key fob, allowing the developers to interface it with any other SPI, I^2C, and UART-based microcontrollers.

The RC522 RFID module works at 13.56 MHz and it communicates with the passive RFID cards at short distances with radio frequency signals (RF). The supplied RFID card has 1 KB (1024 bytes) memory capacity, divided into 16 sectors, where each sector is then divided into 4 blocks. Each block consists of 16 bytes. Therefore, 4 blocks × 16 bytes × 16 sectors = 1024 bytes.

12.2 RC522 RFID reader pins

The RC522 RFID module has the pin layout as shown in Figure 12.2. The pin definitions are:

MISO: SPI output from RC522 when SPI interface is enabled. Serial clock when I²C is enabled. Serial data output when UART is enabled.

MOSI: SPI input to RC522

SCK: SPI clock input

SDA: SPI chip-enabled when SPI enabled. Serial data when I²C-enabled. Serial data input when UART-enabled.

RST: Reset input. When LOW, the module enters power-down mode where the oscillator is turned OFF and the input pins are disconnected.

VCC: +3.3 V power

GND: Power supply ground

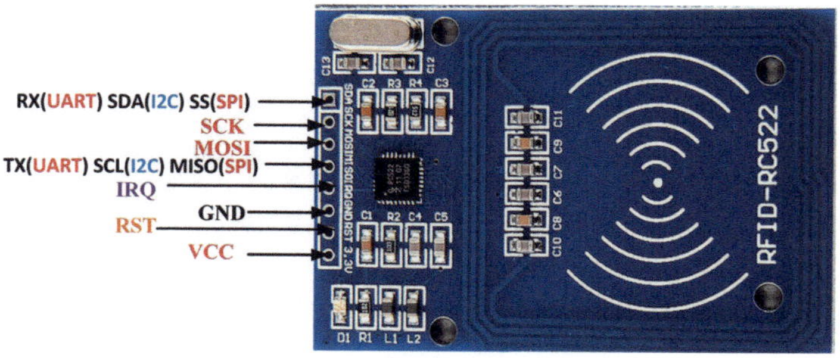

Figure 12.2: RC522 pins.

12.3 Interfacing RC522 RFID reader module to Raspberry Pi Pico W

Before you interface the RC522 module you have to know the SPI bus pins of the Raspberry Pi Pico W which are the same as the standard Pico pins as shown in Figure 12.3. There are two SPI pins available at various ports of the Pico.

Figure 12.3: SPI bus pins of Raspberry Pi Pico W.

The following pins are used to interface the RC522 module to Pico:

RC522 Module	Raspberry Pi Pico W
SDA (CS)	GP1
SCK	GP2
MOSI	GP3
MISO	GP4
GND	GND
RST	GP0
VCC	+3.3 V
IRQ pin is not used	

12.4 Project 1: Finding the Tag ID

Description: In this project, you will display the Tag ID of the supplied RFID Tag on the Serial Monitor.

Block diagram: Figure 12.4 shows the block diagram of the project.

Figure 12.4: Block diagram of the project.

Circuit diagram: The connections between the Pico and the RFID reader are as follows (*be careful not to connect the power pin to +5 V*). Figure 12.5 shows the circuit diagram of the project:

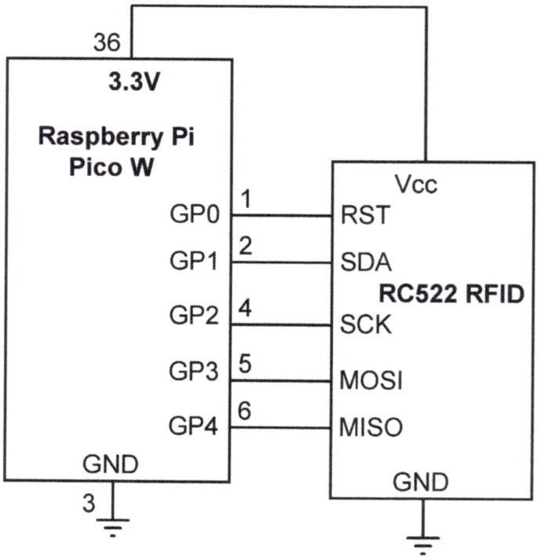

Figure 12.5: Circuit diagram of the project.

Program listing: Before using the RFID reader, you have to add the RFID library to your Pico. The library is called **mfrc522.py** and is available on the Internet. A copy of this library is included in the web site of the book. You must copy this library to your Pico with the exact name **mfrc522.py**.

The program listing is shown in Figure 12.6 (Program: **RFID.py**). At the beginning of the program, **mfrc522** library is imported. The program then displays a message to put the Tag close to the reader. The reader is initialized and a request is sent to the reader. If a Tag is detected, it is selected and the Tag ID is read and displayed on the Thonny screen.

```
#------------------------------------------------------------------------------
#         RFID PROGRAM TO DISPLAY TAG ID
#         ==============================
#
# This program displays the Tag ID of the Tag supplied with the RFID reader
#
# Author: Dogan Ibrahim
# File  : RFID.py
# Date  : October, 2022
#-------------------------------------------------------------------------------
from mfrc522 import MFRC522

reader = MFRC522(spi_id=0,sck=2,miso=4,mosi=3,cs=1,rst=0)

print("Put Tag near the reader...")

readflag = 0
while readflag == 0:
    reader.init()
    status, tag_type = reader.request(reader.REQIDL)

    if status == reader.OK:
        status, uid = reader.SelectTagSN()

        if status == reader.OK:
            card = int.from_bytes(bytes(uid),"little",False)
            print("Your Card ID (Decial): " + str(card))
            print("Your Card ID (HEX):", hex(card))
            readflag = 1
```

Figure 12.6: Program: ***RFID.py****.*

Figure 12.7 shows the Tag ID displayed as a decimal number: 453492048, or hexadecimal number: 0x1b07bd50

```
Put Tag near the reader...
Your Card ID (Decial): 453492048
Your Card ID (HEX): 0x1b07bd50

>>>
```

Figure 12.7: Displaying the Tag ID.

12.5 Project 2: RFID door lock access with relay

Description: In this project, the RFID reader and a relay are both connected to Pico. It is assumed that a secure door entry is operated with a relay and an actuator and is protected with an RFID system. Placing an authorized Tag card near the RFID reader can only open the door. The relay is activated for 15 seconds. After this time, it is deactivated so that the door can be closed.

Block diagram: Figure 12.8 shows the block diagram of the project.

Figure 12.8: Block diagram of the project.

Circuit diagram: The circuit diagram is shown in Figure 12.9. The diagram is basically the same as in Figure 12.5, but here a relay is added to port GP16.

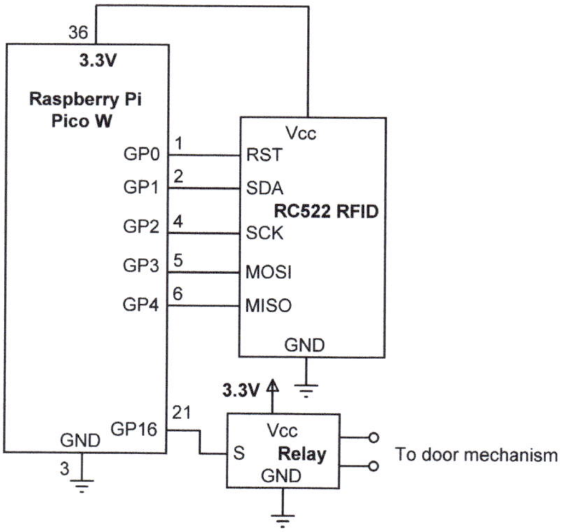

Figure 12.9: Circuit diagram of the project.

Program listing: Figure 12.10 shows the program listing (Program: **RFIDLock**). At the beginning of the program, the **SPI** and **MFRC522** libraries are included in the program. Valid card ID is stored in string **ValidCard**, **RELAY** is assigned to port GP16 and is deactivated. The remaining of the program runs inside the main program loop. Here, the program waits until a Tag is placed near the reader. When a Tag is placed, its ID is read and compared with the **ValidCard**. If there is a match, it is assumed that the Tag is authorized and then **RELAY** is activated for 15 seconds which in turn causes the door to open. If the card is not valid, the relay remains deactivated. This process is repeated forever.

```
#----------------------------------------------------------------------------
#         RFID DOORLOCK ACCESS
#        ====================
#
# This activates a relay if an authorized valid card is placed near the
# RFID reader. The relay remains active for 15 seconds.
#
# Author: Dogan Ibrahim
# File  : RFIDLock.py
# Date  : October, 2022
#------------------------------------------------------------------------------
from mfrc522 import MFRC522
import utime
from machine import Pin

reader = MFRC522(spi_id=0,sck=2,miso=4,mosi=3,cs=1,rst=0)

ValidCard = 453492048                     # Authorized card ID
RELAY = Pin(16, Pin.OUT)                  # RELAY at GP16
RELAY.value(0)                            # Deactivate RELAY

while True:
    reader.init()
    status, tag_type = reader.request(reader.REQIDL)

    if status == reader.OK:
        status, uid = reader.SelectTagSN()

        if status == reader.OK:
            card = int.from_bytes(bytes(uid),"little",False)
            if card == ValidCard:
                RELAY.value(1)            # RELAY ON
                utime.sleep(15)           # Wait 15 seconds
                RELAY.value(0)            # RELAY OFF
```

Figure 12.10: Program: **RFIDLock**.

12.6 Project 3: Multi-tag RFID access system with LCD

Description: In this project, the RFID reader and a relay are both connected to Pico as in the previous project. Additionally, an I^2C-based LCD is connected. The project can handle many users with different Tag IDs. The names of authorized people and their Tag IDs are stored in the program. Only one of these authorized people can activate the relay to open the door.

Block diagram: Figure 12.11 shows the block diagram of the project.

Figure 12.11: Block diagram of the project.

Circuit diagram: The circuit diagram is similar to Figure 12.9, but here additionally an I²C is connected to Pico as shown in Figure 12.12. The LCD is connected to the I2C1 pins while the RFID is connected to the I2C0 pins. Notice that the TXS0102 voltage translator chip is used between the LCD and Pico.

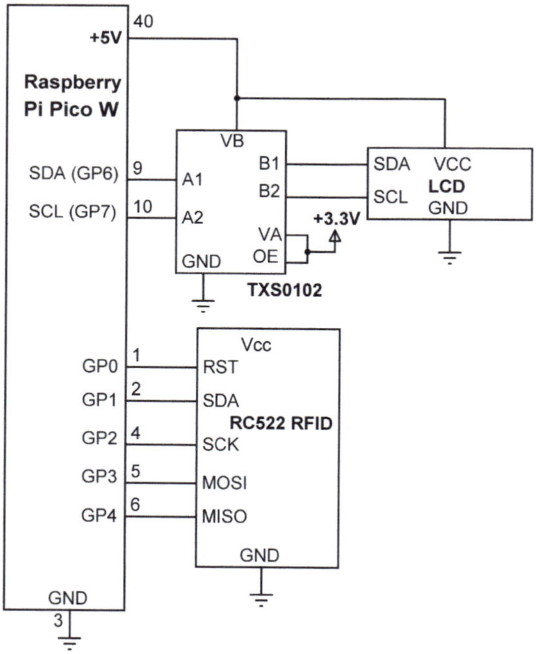

Figure 12.12: Circuit diagram.

Program listing: Figure 12.13 shows the program listing (Program: **members.py**). At the beginning of the program, the LCD and RFID I²C bus are initialized, RELAY is configured at GP16 as an output. Then a dictionary is created with the name **members**. This dictionary stores the Tag ID numbers and the names of the people authorized to have the Tags. Although only five names and Tag IDs are shown in the dictionary, it can easily be expanded. In this project, the Tag ID author is authorized to a person called **Mary Walker**.

```
members={
    123456789: "Jones Smith",
    453492048: "Mary Walker",
    554433765: "Susan Richard",
    545567664: "Mark Beasley",
    455545321: "David Johnson"
}
```

When the program is started, the user is requested to put his/her Tag near the reader and the message **Touch your Tag** is displayed on the LCD (Figure 12.14). The program reads the Tag ID (**card**) and compares with the IDs listed in the dictionary using statement **members.get(card)**. If the user is not authorized, then **None** is returned and the program displays **Not Authorized** on the LCD. The process is repeated after 5 seconds where users are required to touch their Tags. If on the other hand, the Tag is authorized, then text **Authorized** is displayed at the top row of the LCD. The bottom row of the LCD displays the name of the authorized person (Figure 12.15). At the same time, the RELAY is activated for 10 seconds to let the door open. After 10 seconds, the RELAY is deactivated and the process repeats.

```
#-----------------------------------------------------------
#              MULTI-TAG ACCESS CONTROL WITH LCD
#              =================================
#
# In this program names and Tag IDs of authorized users are
# listed in a dictionary. A relay is activated to open a door
# when an authorized user places his/her Tag near the door.
# The name of the authorized user is displayed on the LCD
#
# Author: Dogan Ibrahim
# File  : members.py
# Date  : October 2022
#-------------------------------------------------------------
import machine
from machine import I2C, Pin
from lcd_api import LcdApi
from i2c_lcd import I2cLcd
import utime
from mfrc522 import MFRC522


#
# RFID I2C setup
#
reader = MFRC522(spi_id=0,sck=2,miso=4,mosi=3,cs=1,rst=0)


#
# LCD I2C setup
```

```
#
I2C_ADDR = 0x27                          # I2C LCD address
NRows = 2                                # Number of rows
NColumns = 16                            # Number of columns

i2c = I2C(1, sda=machine.Pin(6), scl=machine.Pin(7), freq=400000)
lcd = I2cLcd(i2c, I2C_ADDR, NRows, NColumns)

RELAY = Pin(16, Pin.OUT)                 # Relay is output
RELAY.value(0)                           # Deactivate relay

#
# Enter valid Tag IDs and authorized member names as in the
# following dictionary. You can extend this table as required
#
members={
    123456789: "Jones Smith",
    453492048: "Mary Walker",
    554433765: "Susan Richard",
    545567664: "Mark Beasley",
    455545321: "David Johnson"
}

lcd.clear()
lcd.putstr("Touch your Tag")

#
# Main program. Read a Tag and check if authorized. If so, display
# the name of the authorized person and activate the relay for 10s
#
while True:
    reader.init()
    status, tag_type = reader.request(reader.REQIDL)
    if status == reader.OK:
        status, uid = reader.SelectTagSN()

        if status == reader.OK:
            card = int.from_bytes(bytes(uid),"little",False)
            auth = members.get(card)
            lcd.clear()
            lcd.move_to(0, 0)
            if auth == None:
                lcd.putstr("Not Authorized")
                utime.sleep(5)
                lcd.clear()
            else:
```

```
            lcd.putstr("Authorized")
            lcd.move_to(0, 1)
            lcd.putstr(auth)
            RELAY.value(1)
            utime.sleep(10)
            RELAY.value(0)
            lcd.clear()
        lcd.putstr("Touch your Tag")
```

Figure 12.13: Program: **members.py**.

Figure 12.14: Displaying text Touch your Tag.

Figure 12.15: Authorized user.

Suggestion for further work: The program can be modified so that the names of the authorized users and their Tag IDs are stored in a file on the Raspberry Pi Pico W.

Index